PHYSICAL THEORY

PHYSICAL THEORY

Method and Interpretation

Edited by Lawrence Sklar

OXFORD
UNIVERSITY PRESS

Oxford University Press is a department of the University of Oxford.
It furthers the University's objective of excellence in research, scholarship,
and education by publishing worldwide.

Oxford   New York
Auckland   Cape Town   Dar es Salaam   Hong Kong   Karachi
Kuala Lumpur   Madrid   Melbourne   Mexico City   Nairobi
New Delhi   Shanghai   Taipei   Toronto

With offices in
Argentina   Austria   Brazil   Chile   Czech Republic   France   Greece
Guatemala   Hungary   Italy   Japan   Poland   Portugal   Singapore
South Korea   Switzerland   Thailand   Turkey   Ukraine   Vietnam

Oxford is a registered trademark of Oxford University Press
in the UK and certain other countries.

Published in the United States of America by
Oxford University Press
198 Madison Avenue, New York, NY 10016

© Oxford University Press 2015

All rights reserved. No part of this publication may be reproduced, stored in a
retrieval system, or transmitted, in any form or by any means, without the prior
permission in writing of Oxford University Press, or as expressly permitted by law,
by license, or under terms agreed with the appropriate reproduction rights organization.
Inquiries concerning reproduction outside the scope of the above should be sent to the
Rights Department, Oxford University Press, at the address above.

You must not circulate this work in any other form
and you must impose this same condition on any acquirer.

Library of Congress Cataloging-in-Publication Data
Physical theory : method and interpretation / edited by Lawrence Sklar.
  pages cm
Includes index.
ISBN 978–0–19–514564–9 (hardback : alk. paper)   1. Science—
-Methodology.   2. Science—Philosophy.   I. Sklar, Lawrence, editor of compilation.
Q175.P514 2014
501—dc23
2013010048

1 3 5 7 9 8 6 4 2
Printed in the United States of America
on acid-free paper

CONTENTS

List of Contributors — vii

Introduction — 1

PART ONE
SCIENTIFIC METHOD

1. Scientific Explanation — 9
   James Woodward

2. Probabilistic Explanation — 40
   Michael Strevens

3. Laws of Nature — 63
   Marc Lange

CONTENTS

4. Reading Nature: Realist, Instrumentalist, and
   Quietist Interpretations of Scientific Theories   94
   P. Kyle Stanford

5. Structure and Logic   127
   Simon Saunders and Kerry McKenzie

6. Evolution and Revolution in Science   163
   Jarrett Leplin

PART TWO
FOUNDATIONS OF PHYSICS

7. What Can We Learn about the Ontology of Space
   and Time from the Theory of Relativity?   185
   John D. Norton

8. QM_∞ 229
   Laura Ruetsche

9. Statistical Mechanics in Physical Theory   269
   Lawrence Sklar

Index   285

LIST OF CONTRIBUTORS

Marc Lange University of North Carolina at Chapel Hill

Jarrett Leplin University of North Carolina at Greensboro

Kerry McKenzie University of Calgary

John D. Norton University of Pittsburgh

Laura Ruetsche University of Michigan

Simon Saunders Oxford University

Lawrence Sklar University of Michigan

P. Kyle Stanford University of California, Irvine

Michael Strevens New York University

James Woodward University of Pittsburgh

Introduction

The contributions to this volume provide a survey of two aspects of contemporary philosophy of science. The pieces in the first part of the volume give the reader a rich sampling of current work in the general methodology of science. The pieces in the second part of the volume offer a sample of the current work exploring the foundations of our most general and basic science, contemporary physics.

1. SCIENTIFIC METHOD

Three general problem areas have dominated discussion about the methods of the sciences. How do scientists answer "why?" questions—that is, what kinds of explanations do the sciences offer of the phenomena in their domains? What is the nature of the broad structures we call "theories"—how do theories unify science and provide a context for particular explanations? Finally, what is our rationale for believing or disbelieving in the explanations and theories proffered by our current, best available science? These problem areas are explored by the contributors to part 1 of the volume.

In "Scientific Explanation," James Woodward explores what the structure of an explanatory answer to a "why?" question might be in science. He outlines the important idea that to explain is to place what occurs under a generalization, exceptionless or statistical, about what kinds of things occur. Then he notes a number of objections to that simple notion of explanation. Woodward then explores some proposals to supplement this subsumption model of explanation, including the proposal to demand that explanations in science be unifying and the proposal that a basic notion of causation is essential to our concept of explanation.

INTRODUCTION

In "Probabilistic Explanation," Michael Strevens pursues the structure of explanations that are statistical or probabilistic further. The subsumption model of statistical explanation along with some of its demands (such as maximal specificity) and results (such as the epistemic relativity of explanations) are explored. Proposals that causation is essential even to statistical explanation and that statistical explanations are founded on irreducible stochastic dispositions are explored, as is the issue of whether or not one can have genuine statistical explanations in a deterministic world. Finally, Strevens suggests that a notion of "robustness" is a requirement of a probabilistic explanation.

When scientists proffer explanations, they refer us to relevant generalizations, exceptionless or statistical, that they call "laws of nature." But what is a law of nature? This is the subject of Marc Lange's "Laws of Nature." Lange takes up the problems of the distinction between "genuine laws" and "mere accidental generalizations" and of the relation of a law to the counterfactual conditionals it supports. The notion of "natural necessities" is discussed, as is the idea that what makes us consider a generalization a law is the place of that generalization in our scientific hierarchy of principles. Finally, the role of laws in the "inexact sciences" is discussed.

Scientists do not stop at the level of laws, however. The laws are fitted into that larger structure of science we call "theories." One aspect of theories is that they often posit new entities and properties, and often these entities and properties are not within the realm of the "observable" as that is naively understood. How should we understand what the scientist is really claiming when some such theory is proposed? These are the issues taken up by P. Kyle Stanford in "Reading Nature: Realist, Instrumentalist, and Quietist Interpretations of Scientific Theories." Should we understand the theories to be telling us of the real existence of a world beyond our powers of observation, a world that explains why the observable things behave the way they do, or should we rather think of theories as useful "instruments" for predicting correlations among the observables but not as positing brand new realms of being? Can we even make such a distinction in a legitimate way? The epistemological, semantic, and ontological aspects of those questions are taken up in this contribution.

The nature of theories is taken up from a different perspective by Simon Saunders and Kerry McKenzie in "Structure and Logic." One proposal for understanding how theories work takes the theory to consist of an observational base with a theoretical "structure" imposed on it. A deep understanding of what this might mean was suggested by Frank Ramsey many years ago. But how are we to understand the "Ramsey sentence" reconstruction of a theory? Does the theoretical structure guide us to real things and features in the world or only to a structural "embedding" for the genuine, observational structure. Claims concerning the "underdetermination" of the referential structure of the theory are often taken as conclusive arguments against a realist reading of the theory. But is this correct? These are some of the issues discussed in this contribution.

Finally, in this methodological part of this volume there are the issues of the grounds on which we are justified in accepting or rejecting a scientific theory. Some crucial issues here are discussed by Jarrett Leplin in "Evolution and Revolution in Science." A profound problem that has been much discussed is whether the changes in science are gradual and "progressive" or whether instead they are "revolutionary," with no clear conceptual path from theory to later theory that we can think of as taking us ever closer to some final truth. Here Leplin takes the notions of "evolution" versus "revolution" seriously as they are applied outside philosophy of science to explore how adequate these terms can be in describing scientific change.

2. FOUNDATIONS OF PHYSICS

Science comes in a wide variety of forms, and each particular science raises its own philosophical questions. In the social sciences we find concepts and explanations that look quite different from the mathematical specifications of states and formal presentations of laws that we find in physics. In biology we find the idiosyncratic concepts and explanations that frame the theory of evolution.

One particularly important and philosophically puzzling branch of science is fundamental physics. Here we find the theories that are alleged

INTRODUCTION

to be the broadest and in many ways the deepest in all the sciences. It is often claimed as well that our assurance of the "correctness" of these theories is much less controversial than is our epistemic certitude in other science. Despite this, these theories of foundational physics give rise to the most intractable and long-lived philosophical quandaries in all philosophy of science. These foundational theories introduce novel concepts and ontologies remote from those of everyday language and experience. The explanations these theories offer are frequently deeply puzzling as well. Special kinds of probabilistic explanation are introduced. The theories seem to posit "actions at a distance" that are hard to reconcile with our ideas that causality is local. Many explanations indeed do not seem "causal" in any ordinary sense at all. It is to these special problems arising out of the curious nature of these fundamental physical theories that the contributions in the second part of this volume are directed.

In "What Can We Learn about the Ontology of Space and Time from the Theory of Relativity?" John D. Norton takes up the issue of just what exactly the theories of special and general relativity tell us about space and time that forces us to revise the ideas we had about these two "frameworks" of all phenomena in our common sense and in prerelativistic physics. What are the features that are genuinely novel, that follow from the new theories, that are part and parcel of a literal understanding of the theories, and that are robust in the sense of not being faced with contradictory "morals" that can also be drawn from these new theories?

Quantum mechanics notoriously introduced deep puzzles about what it was telling us about the world. Quantum field theory, the generalization of quantum mechanics developed to deal with the creation and destruction of elementary particles, introduced its own new interpretative difficulties. These are explored by Laura Ruetsche in "QM_∞." In quantum mechanics the so-called commutation operators on the basic descriptive operators fix the descriptive realm of the theory. How are we to understand this theory that seems to allow for distinct, incompatible ways of representing the basic features of the world? Ruetsche explores some of the answers given to this deep problem.

The first theory of physics in which probabilistic notions played a constitutive role was statistical mechanics, the attempt to ground

INTRODUCTION

the macroscopic theory of heat and temperature on the underlying atomic constitution of matter. This theory is explored in Lawrence Sklar's "Statistical Mechanics in Physical Theory." Here such issues as the need for and structure of the basic probabilistic posits of the theory, the justification of these posits, and the issues concerning how a time-asymmetrical theory of heat could be consistent with a time-symmetrical underlying dynamics of the constituents of matter are discussed. The claim that our very idea that time is asymmetrical rests on these physical asymmetries is also explored.

PART I

SCIENTIFIC METHOD

Chapter 1

Scientific Explanation

JAMES WOODWARD

1. INTRODUCTION

Accounts of scientific explanation have been a major focus of discussion in philosophy of science for many decades. It is a presupposition of such accounts that science aims at (and sometimes succeeds in) providing "explanations" and that this activity contrasts in a nontrivial way with other aspects of the scientific enterprise that are not explanatory ("mere description" is one obvious candidate for such an alternative). If this were not true, there would be no distinctive topic of scientific explanation. Further, as a normative matter the philosophical literature assumes that (at least usually or other things being equal) it is a desirable feature in a theory that it provide successful explanations—explanation is seen as a goal of scientific inquiry.

Of course the notion of explanation, as it occurs in ordinary speech, encompasses many different possibilities (one speaks of explaining the meaning or a word, how to solve a differential equation, and so on). However, with some conspicuous exceptions (e.g., Scriven 1962), the literature on scientific explanation has tended to assume a narrower focus: very roughly, the task is conceived as providing an account or "model" of the notion of explanation that is in play when one speaks of explaining why some particular event or regular pattern of events occurs. Examples include the explanation of Kepler's laws in terms of Newtonian mechanics, the explanation of phenomena like refraction and diffraction in terms of the wave theory of light, the explanation of the distribution of phenotypic characters in subsequent generations in terms of the genotypes of parents and the principles of Mendelian genetics, and the explanation of why some particular person or group of people developed a certain disease.

In addition, most of the influential accounts of explanation tend to assume that there is some common abstract structure shared by successful explanations in all areas of science—physics, biology, psychology—and perhaps in common sense as well. The task of a model of explanation is to characterize this common structure in a way that abstracts from the specific content of (or the specific empirical assumptions that underlie) the different scientific disciplines. Relatedly, advocates of the various competing models discussed below have also tended to assume that explanation in the sense they are trying to capture is a relatively "objective," audience-independent matter—that is, that the right account of explanation is not just the banality that different people will find different pieces of information explanatory, depending on their interests and background beliefs. This assumption is of course closely connected to the idea that successful explanation is a normatively appealing goal of science—it is commonly supposed that a highly contextual and audience-dependent account of explanation is not a good candidate for this normative role.

2. THE DN MODEL

Contemporary discussion of scientific explanation in philosophy of science has been greatly shaped by the formulation of the deductive-nomological (DN) model in the middle part of the twentieth century. Although there are many anticipations and alternative statements of the basic idea, the most detailed and influential version is that of Carl Hempel (1965b). Assume that we are dealing with a domain in which the available laws are deterministic. Then, according to the DN model, scientific explanation has the form of a valid deductive argument in which the fact to be explained (the *explanandum* in Hempel's terminology) is deduced from other premises that do the explaining (the *explanans*). (This is the *deductive* part of the DN model.) These premises must be true and must include at least one "law of nature" that figures in the deduction in a nonredundant way in the sense that the deduction would no longer be valid if this premise were removed. The law premise is the *nomological* part of the model, "nomological" being a philosophical term of art for "lawful." Commonly, the *explanans* will include other, nonlawful premises as well—for example, these may include statements about "initial conditions" and other matters of particular fact. The *explanandum* may be either a generalization or itself a claim about a particular matter of fact, although Hempel tended to focus largely on the latter possibility.

As an illustration, consider the derivation of the position of a planet at some future time t from (a) information about its earlier position and momentum and the position, velocity, and mass of other gravitating bodies (other planets, the sun) and (b) Newton's laws of motion and the Newtonian gravitational inverse square law. The information in (a) represents initial conditions, and in (b) are laws of nature that figure nonredundantly in the deduction. The resulting derivation is a sound, deductively valid argument that constitutes a DN explanation of the position of the planet.

3. THE IS MODEL

Hempel recognized that in many areas of science generalizations are statistical rather than deterministic in form. When such generalizations take the form of statistical "laws," Hempel (1965a, 376–412) suggested that we may think of them as providing explanations of individual outcomes in accordance with a distinctive form of explanation that he calls inductive-statistical (IS) explanation. Formulating an adequate model of statistical explanation turned out to be a complicated matter and gave rise to a substantial literature, but Hempel's underlying idea is straightforward: statistical laws explain individual outcomes to the extent that they show that, given the prevailing initial conditions, those outcomes are highly probable. For example, if it is a statistical law that a particular radium atom has a probability of 0.9 of decaying within a certain time interval Δt and the atom does decay in this time interval, then we can provide an IS explanation of the decay by citing these facts. Similarly, if it is a statistical law that those with staph infection have a probability of 0.8 of recovering when penicillin is administered (and certain other conditions are met), we can use this law to provide an IS explanation of why some particular patient, Jones, recovers from such an infection.

4. THE ROLE OF LAWS

The requirement that DN explanations be deductively valid seems clear and straightforward, but how exactly should we understand the requirement that the deduction contains a law of nature? Hempel thought of laws as (a proper subset of those) generalizations describing "regularities" or uniform patterns of occurrence in nature. In the canonical case, these generalizations have the form of universally quantified conditionals (for any object x, if x is an F, then x is also a G—that is, all Fs are Gs). A DN explanation of why a is G thus will appeal to this nomological premise and to the claim that a is F. However, as Hempel and other defenders of the DN model recognized, not all true generalizations having this form are naturally regarded as laws; instead some appear to be only "accidentally" true. To use Hempel's (1965a) example, the generalization "All members of the Greenbury schoolboard for 1964 are bald" (339) seems intuitively to be, even if true, accidentally so rather than a law. If so, this generalization cannot be used to provide a DN explanation of why some individual member of that board is bald—an assessment that certainly conforms to our intuitive judgment about this case.

As Hempel recognized, it would be desirable not to leave judgments about whether a true generalization counts as a law or not at an intuitive level but instead to provide clear, noncircular criteria that allow us to distinguish laws and nonlaws in a more principled manner. Various candidates for these distinguishing criteria are considered by Hempel (1965a, 335ff.): for example, the suggestions that laws (as opposed to true generalizations) must contain purely qualitative predicates, that they must contain "projectable" predicates, and that they must support counterfactuals. Hempel argues, however, that these candidates either fail to distinguish between laws and accidental regularities or are in other respects unsatisfactory. Although there has been a great deal of subsequent discussion in the intervening decades and an approach to the problem that enjoys considerable support (the so-called Mill-Ramsey-Lewis [MRL] theory, discussed briefly in section 11), philosophers are still far from a consensus regarding the law-nonlaw distinction. Indeed there is considerable disagreement even about the role laws play in science, both in general and in the various particular sciences, with some philosophers defending a "no-law" view, according to which there is nothing corresponding to the philosophical notion of a law of nature in any area of science, including physics and chemistry, and others claiming that even if there are laws in physics and chemistry, there are few or no laws in disciplines like biology and psychology.[1] The development of a more adequate account that allows us to distinguish between laws and other true generalizations (and which gives us some insight into the role laws play in explanation) thus remains an important project for defenders of the DN and other law-based models of explanation.

In practice, defenders of the DN model have often assumed an informal and very permissive notion of law, according to which "laws" include not just such fundamental physical principles as the Schrödinger equation but far more local and exception-ridden generalizations from the special sciences (e.g., Mendel's "laws" of segregation and independent assortment) as well as garden-variety causal generalizations, such as "Aspirin cures headaches" and "If a rock strikes a window hard enough, the window will shatter." Thus even this last generalization is often regarded as capable of serving as a nomological premise in a DN explanation of some particular episode of window shattering. This permissive conception of law in turn is connected to the close connection between the DN model and Humean or regularity-based accounts of causation, which is discussed below.

---

1. For general disagreements about the role played by laws in science, compare Hempel 1965a and Earman 1993 with van Fraassen 1989 and Giere 1999.

5. THE DN/IS MODEL IN CONTEXT

Many interesting historical questions about the DN/IS model remain largely unexplored. Why did "scientific explanation" emerge when it did as a major topic for philosophical discussion? Why were the "logical empiricist" philosophers of science who defended the DN model so willing to accept the idea that science provides "explanations," given the tendency of many earlier writers in the positivist tradition to think of "explanation" as a rather subjective or "metaphysical" matter and to contrast it unfavorably with "description" (or "simple and economical description"), which they regarded as a more legitimate goal for empirical science? And why was discussion, at least initially, organized around "explanation" rather than "causation," since (as we shall observe) it is often the latter notion that seems to be of central interest in subsequent debates and since the former notion seems (to contemporary sensibilities) somewhat squishy and ill-defined?

At least part of the answer to this last question seems to be that Hempel and other defenders of the DN model inherited standard empiricist or Humean scruples about the notion of causation. They assumed that causal notions are only (scientifically or metaphysically) acceptable to the extent that it is possible to paraphrase or redescribe them in ways that satisfied empiricist criteria for meaningfulness and legitimacy. One obvious way of doing this was to take causal claims to be tantamount to claims about the obtaining of regularities, and it is just this idea that is captured by the DN/IS model (see below). Part of the initial appeal of the topic of "scientific explanation" was thus that it functioned as a more respectable surrogate for (or entry point into) the problematic topic of causation.[2] Another motivation was the interest of Hempel and other early defenders of the DN model in forms of explanation, such as "functional explanation" (thought to be employed in such special sciences as biology and anthropology), that were not obviously causal. This also

---

2. See Cartwright 2004 for a similar diagnosis and for another survey of some of the issues described here. Sklar 1999 is a very interesting discussion of the historical background and motivation for the DN model.

made it natural to frame discussion around a broad category of explanation (see Hempel 1965c).

6. MOTIVATION FOR THE DN/IS MODELS

Why think that successful explanation must have a DN or an IS structure? Hempel appeals to two interrelated ideas. The first has to do with the point or goal of explanation. A DN/IS explanation shows that the phenomenon to be explained "was to be expected" on the basis of a law, and "it is in this sense that the explanation enables us to understand why the phenomenon occurred" (Hempel 1965a, 337). This connection between explanation and law-based (or "nomological") expectability has considerable intuitive appeal; one thing we might hope for from an explanation is that it diminish our feeling that the explanandum phenomenon is surprising, arbitrary, or unexpected, and showing that the explanandum follows from a law and other conditions, either for certain or with high probability, contributes to accomplishing this. In addition, it is widely (although not universally, see footnote 1) believed that in some fundamental areas of science, like physics, the discovery of laws of nature is centrally important. The DN/IS model resonates with this observation by assigning laws a central role in explanation. We should note, however, that these remarks leave some obvious questions unaddressed. What is so special, from the point of view of explanation, about *nomological* expectability? After all, there are other ways, including appealing to true accidental generalizations, of showing that an outcome "was to be expected." Why are not some of these also sufficient for explanation? What exactly is it that citing a law contributes to successful explanation? And even if a demonstration of nomological expectability is necessary for explanation, how do we know that there are not other requirements that a successful explanation must meet as well?

A second and closely related idea that motivates the DN/IS model has to do with an assumed connection between causation and the instantiation of laws or regularities—what we might call the assumption of the nomological (which in this context means regularity-based) character of causation. Hempel is of course aware that we often think of explaining an outcome as a matter of providing information about its causes. However, for the reasons alluded to in section 5, he is unwilling to take the notion of causation as primitive or unanalyzed in the theory of explanation. Instead he insists that causal claims always "implicitly claim" or "presuppose" the existence of some associated law or laws, according to which the candidate for cause is part of some larger complex of "antecedent conditions" that are linked via a regularity to the explanandum phenomenon. The correct account of causation is thus assumed to be broadly Humean in the sense that causal claims are to be explicated in terms of the obtaining of regularities. Generalizations describing these regularities in turn serve as nomological premises in a DN (or perhaps an IS) explanation of the effect that we wish to explain, so that all causal explanation turns out to be, at least implicitly, DN or IS explanation. The role assigned to nomic expectability thus fits naturally with David Hume's story about how experience of the regular association leads to our learning to expect or anticipate the effect when we observe the cause (cf. Cartwright 2004).

7. COUNTEREXAMPLES TO THE DN/IS MODEL

A number of well-known "counterexamples" have been advanced against both the sufficiency (counterexamples [7.1] and [7.2] below) and the necessity (counterexample [7.3] below) of the DN/IS requirements on explanation. These are commonly presented as examples in which our "intuitive" or "preanalytic" judgments about whether explanations have been provided seem to differ from the judgments dictated by the DN/IS model. The common thread running through these counterexamples has to do with the apparent failure of the DN/IS model to adequately capture how causal notions enter into our judgments about explanation—that is, the counterexamples appear to be a reflection of the inadequacy of the simple version of a regularity theory of causation assumed in the DN/IS model.

7.1. Counterexample (7.1)

Many explanations exhibit "directional" or asymmetrical features that do not seem to be captured by the DN/IS model. From information about the length l of a simple pendulum and the value of the acceleration g produced by gravity, one may derive its period T by employing the "law" $T = 2\pi\sqrt{l/g}$. This derivation satisfies the DN requirements and seems intuitively to be explanatory. However, by running the derivation in the "opposite" direction, one may deduce the length l from the values of T and g and the generalization $l = gT^2 / 4\pi^2$. This derivation again satisfies the DN requirements assuming that $l = gT^2 / 4\pi^2$ is also a law but seems intuitively to be no explanation of why the pendulum has the length that it does (cf. Bromberger 1966). Other illustrations of the same basic point are readily produced. From the mass and acceleration of a test particle and the law $F_t = ma$, one can deduce the total force F_t incident on the particle, but intuitively this is no explanation of why the particle experiences the force F_t. Instead the explanation for this is to be sought in the various component forces that sum to produce F_t, the specialized force laws governing these, and so on.

7.2. Counterexample (7.2)

The presence of certain kinds of irrelevant information seems to undermine the goodness of explanations, even if these satisfy the DN requirements. To employ a famous example (Salmon 1971), from the generalization (M) "All males who take birth control pills fail to get pregnant" and the additional premise that Jones is a male who takes birth control pills, one can deduce that Jones fails to get pregnant. Arguably (M) counts as a law according to the usual philosophical criteria employed by philosophers. However, the resulting derivation seems not to explain why Jones fails to get pregnant. Intuitively, this is because, given that Jones is male, his taking birth control pills is irrelevant to whether he gets pregnant. As this example illustrates, a condition that is nomologically sufficient for an outcome need not be explanatorily relevant to the outcome. Successful explanation seems to require the citing of *relevant* conditions (cf. Salmon 1971).

7.3. Counterexample (7.3)

Suppose that only those who have latent syphilis (s) develop paresis (p) but that the probability of p, given s, is low, say, 0.3 (cf. Scriven 1959). If Jones develops paresis, it seems (again intuitively) explanatory to cite the fact that he has s. But in doing so we have not (at least explicitly) cited laws and conditions that make p certain or even highly probable, which is what the DN/IS model demands.

The reaction of many philosophers to these counterexamples has been that they show that the DN/IS model fails to adequately capture the role of distinctively *causal* information in explanation.[3] Thus the directional features that seem to be omitted from the DN/IS model appear to be closely connected to (if not identical with) the asymmetry of causation: the length of the pendulum is one of the causes of it having the period that it does and not vice versa. It is because it is legitimate to explain effects in terms of their causes but not vice versa that it is appropriate to explain the period in terms of the length and not vice versa. Counterexample (7.2) seems to trade on the fact point that (barring complications having to do with causal preemption and overdetermination) causes must make a difference to their effects. Taking birth control pills does not make a difference to whether males become pregnant (they will not become pregnant whether or not they take birth control pills) and in consequence does not explain this outcome. More generally, counterexample (7.2) shows that a factor can be (or can be part of) a nomologically sufficient condition for an outcome and yet not cause it. With respect to counterexample (7.3), we seem willing to accept it as explanatory (to the extent that we do), because it is natural to think of latent syphilis as a cause of paresis. Counterexample (7.3) suggests that a factor can cause an outcome (and hence, arguably, explain it) without being nomologically sufficient for it (or even rendering it probable).

Assuming this analysis is correct, it appears that the way forward (or at least a large part of the way forward) is to focus more directly on the role of causal considerations in explanation and on the development of

---

3. See Salmon 1989; Woodward 2003; and Cartwright 2004, among others, for this diagnosis.

a more adequate theory of causation. To a large extent (although by no means entirely) this is the path taken in subsequent philosophical discussion—particularly by Wesley Salmon in his statistical relevance and (subsequently) causal mechanical models of explanation.

8. THE SR MODEL

Salmon's (1971) statistical relevance (SR) model departed from the DN/IS model in a number of major respects. Most fundamentally, the SR model employed a very different formal/mathematical framework—probability theory rather than the first order logic on which DN theorists relied. In this respect it followed a general trend in philosophy of science, which had become manifest by the late sixties and early seventies, of employing ideas from probability theory (rather than logic) to explicate important concepts—evidence, confirmation, cause, explanation, and so on. The SR model also drew on a very different guiding idea about explanation from the idea on which DN theorists relied—this being that explanation requires furnishing causal or explanatorily *relevant* information about the explanandum and that (as counterexamples [7.1] and [7.2] above illustrate) this is *not* just a matter of exhibiting a condition that is nomologically sufficient for (or that probabilifies) the explanandum. Probability-related notions, such as independence/dependence and correlation, turn out to be more successful (even if not entirely satisfactory) at capturing the notion of relevance than logic-based notions.

The details of the SR model are complex (and, in any case, available elsewhere),[4] but the underlying strategy is to assume that the variables that figure in the explanation are random variables with a well-defined joint probability distribution given by a probability function P and then to try to capture the relevance relationships that are essential to explanation by means of so-called *statistical relevance* and *screening off* (conditional independence and dependence) relationships. A factor A is said to be statistically relevant to another factor B in circumstances C if and

---

[4]. In addition to Salmon 1971, see Salmon 1984 and 1989 for exposition of the SR model.

only if either $P(B/A.C) \neq P(B/C)$ or $P(B/\text{-}A.C) \neq P(B/C)$. D screens off A from B if and only if $P(B/A.C.D) = P(B/D.C)$. Very roughly, the intuition underlying the SR model is that we explain an outcome by assembling information about factors that remain statistically relevant to it as we conditionalize on other suitable factors,[5] together with the statistical relevance relationships themselves, as reflected in the underlying probability distribution. We do not include statistically irrelevant factors.[6] Moreover, according to Salmon, such information about statistical relevance relationships tells us about the causes of phenomena we wish to explain. Thus although the SR model presented as a theory of *explanation*, it in effect assumes a theory of *causation* according to which causal relationships can be fully captured by or reduced to facts about relationships between conditional probabilities. In this respect Salmon's assumed theory of causation continues to satisfy broadly Humean constraints, even though it is different from the account of causation tacitly assumed in the DN model.

To illustrate, return to the birth control pills of counterexample (7.2). On Salmon's analysis, whether or not Jones takes birth control pills is not statistically relevant to (and hence, according to the SR model, does not explain) whether he becomes pregnant, because $P($Fails to Become Pregnant / Is Male. Takes Birth Control Pills$) = P($Fails to Become Pregnant / Is Male$)$ and $P($Fails to Become Pregnant / Is Male. Does Not Take Birth Control Pills$) = P($Fails to Become Pregnant / Is Male$)$. These statistical relationships capture or correspond to the fact that taking birth control pills does not cause failure to become pregnant in males. By contrast, $P($Paresis / Latent Syphilis$) \neq P($Paresis$)$, and this would presumably remain true as we conditionalize on other appropriate factors W; hence latent syphilis remains relevant to (and causes and explains) paresis. In the SR model, in contrast to the DN/IS model, it is thus possible to explain low probability events by exhibiting factors that are statistically relevant to their occurrence.

---

5. For an attempt to characterize what "suitable" means, see Salmon 1984.
6. The full details of the SR model are considerably more complex. See Salmon 1984, 36–37.

The SR model contains a number of important insights about explanation, including an appreciation of the central importance of the notion of causal/explanatory relevance. However, there is an important respect in which the model is inadequate: as philosophers (including Salmon himself) soon recognized, it is not possible to fully capture causal relationships in terms of statistical relevance relationships. Instead the causal relationships among a set of variables are greatly underdetermined even by full information about the statistical relevance relationships among them. To the extent that it is true, as the SR model assumes, that explaining an outcome involves providing information about its causes, the model rests on an inadequate theory of causation.

The failure of statistical relevance relations to fully capture causal relationships is reflected in the familiar adage that "correlation is not causation." As an illustration, consider the following three (different) causal structures: in (8.1) C is a common cause of both A and B, with no other causal connections present; in (8.2) A causes C, which in turn causes B; in (8.3) B causes C, which causes A. (Here A, B, and C are random variables rather than names for factors or properties.) If we make assumptions like Salmon's about the relationship between causation and screening off relations, then all three structures imply exactly the same independence and dependence relationships. In all three structures A and B are unconditionally dependent but become independent conditional on C, A and C and B and C are dependent, and so forth. In some cases, it may be possible to appeal to some other considerations (e.g., temporal order) to distinguish among (8.1), (8.2), and (8.3), but statistical relevance relationships alone will not accomplish this.[7] Moreover, there are good reasons to think that even when statistical relevance relationships are supplemented by temporal considerations, they will not always allow us to distinguish among alternative causal structures.

---

7. For additional relevant discussion in the context of probabilistic theories of causation, see Cartwright 1983. Nancy Cartwright was one of the first philosophers to explicitly recognize the underdetermination of causal facts by statistical information. Formal results about the extent of this underdetermination are in Spirtes et al. 2000.

9. THE CM MODEL

In later work Salmon (e.g., 1984) in effect recognized this and devised a new account of explanation—the causal/mechanical (CM) model—that attempts to capture the "something more" that is involved in causation besides mere statistical relevance relationships. One of the key components of the CM model is the notion of a *causal process*. This is a physical process, such as the movement of a billiard ball or light wave through space, that has the ability to "transmit a mark." This means that if the process is altered in some appropriate way (e.g., white light is passed through a red filter), this alteration will persist in the absence an additional external interference. More generally, causal processes have the ability to propagate their own structures from place to place and over time, in a spatiotemporally continuous way, without the need for further outside interactions. Typically, if not always, this involves the transfer of energy and momentum between successive stages of the process. Causal processes contrast with *pseudoprocesses*, which lack these characteristics. The successive positions of a spot of light on the surface of a dome that is cast by a rotating search light represent a pseudoprocess. If we "mark" the light at one position by temporarily interposing a red-colored filter between the light source and the surface, the spot of light will be colored at that point, but this mark will not persist for successive positions of the light spot unless we continuously move the filter in the appropriate way. The spot thus lacks the ability to transmit its own structure without outside supplementation. One distinguishing feature of causal processes is that they are subject to an upper limit on their velocity of propagation—c, the speed of light. By contrast, pseudoprocesses may move at arbitrarily high velocities.

A *causal interaction* occurs when two causal processes (spatiotemporally) intersect and modify each other, as when a collision between two billiard balls results in a change in momentum of both. In general causal processes but not pseudoprocesses are the carriers of causal influence. For the CM model, the difference among the structures (8.1), (8.2), and (8.3) is thus that they involve different causal processes, something that might be revealed in their different implications for the persistence of marks or in different patterns of spatiotemporal connectedness, despite their satisfying the same statistical relevance relationships. It is in this way that the CM model attempts to capture the "something more" that is involved in causation, over and above statistical relevance relationships.

According to the CM model, an explanation of some phenomenon E involves tracing the causal processes and interactions (or some portion of these) that lead up to E. In still more recent work Salmon (e.g., 1994, 1997) retained this basic picture but attempted to characterize the notion of a causal process in terms of conservation laws rather than in terms of markability, an approach to causation that has been extended by others working in the physical process view of causation, such as Philip Dowe (2000).

The CM model represents an attempt to characterize causation in, as it were, physical or material (or as Salmon says, "ontic") terms rather than in terms of the more formal or mathematical relations emphasized in the DN/IS and SR models. It purports to be a model of causation as it exists and operates in our world rather than a model that aims to characterize what causation must involve in all logically possible worlds. (Perhaps in some logically possible worlds causation does not involve spatiotemporally continuous processes and the transfer of energy and momentum, but it does in ours.) The paradigmatic application of the CM model is simple mechanical systems, like colliding billiard balls, in which causal influence is transmitted by spatiotemporal contact and which involve the transfer of quantities like momentum and energy that are locally conserved. The model nicely captures the sense that many people have that there is something especially intelligible or explanatorily satisfying about such mechanical interactions (and the theories that describe them) and something fundamentally unsatisfying from the point of view of explanation about theories that postulate action at a spatiotemporal distance, nonlocal causal influences, and so on. (On the other hand, as we note below, it is also arguably a limitation of the model that its application appears to be limited to theories that postulate causal relations that do not explicitly involve spatio-temporally continuous processes.) Moreover, regardless of what one thinks about the adequacy of Salmon's various characterizations of causal processes, it also seems uncontroversial that the contrast between causal and pseudoprocesses is a real one, that it has considerable scientific importance, and that when the notion of tracing causal processes is applicable (see below), these have, in comparison with pseudoprocesses, a privileged role in explanation. Finally, the CM model deserves credit for drawing attention to the importance of the notion of "mechanism" in explanation—a notion that until recently has been unexplored by philosophers of science, despite the fact that in many areas of science, including in particular the biomedical sciences, explaining a phenomenon is often seen as a matter of elucidating the operation of the mechanisms that produce that outcome.[8] (It is of course a further question whether the CM model provides a useful account of what a mechanism is in contexts outside physics.)

Despite these attractions, the CM model, like its predecessors, suffers from some serious limitations. First and rather ironically, given that it was Salmon who first emphasized the importance of this notion, the CM model fails to adequately capture the role of causal relevance in explanation. The reader is referred to Hitchcock 1995 and Woodward 2003 for details, but the basic point is quite simple: information about the presence of a connecting process between C and E need not tell us anything about how, if at all, changes in C would affect changes in E, and it is this latter information that is crucial to assessments of relevance. For example, when an ordinary rubber ball is thrown against a brick wall and then bounces off, there is a connecting causal process from the thrower to the wall, but the existence of this process is not (in ordinary circumstances) causally or explanatorily relevant to whether the wall continues to stand up (Hausman 2002). This is because (again in ordinary circumstances) a thrown ball will not transfer enough momentum to the wall to knock it down, so whether or not it is thrown makes no difference to this feature of the wall. In other words, whether or not the wall remains standing depends on the magnitude of the momentum transferred, as becomes apparent if we think instead of the impact of a high-velocity cannonball. The information that the trajectory of the ball is a causal process does not convey this.

---

8. For recent discussions of the notion of a mechanism, see Machamer et al. 2000 and Bechtel and Abrahamsen 2005.

A second issue concerns the scope of the CM model. As we have noted, the most straightforward application of the CM model is to simple mechanical systems involving macroscopic or classically behaved objects. The implications of the model for other sorts of systems, those studied both in physics and elsewhere, are less straightforward and arguably less intuitive. Consider an action at a distance theory like Newtonian gravitational theory as originally formulated by Newton. This theory does not trace continuous causal processes and interactions. A literal reading of the CM model thus seems to imply that this theory is entirely unexplanatory, despite its many other virtues. Or consider standard (nonrelativistic) quantum mechanical treatments of such phenomena as nonclassical electron tunneling through a potential barrier. When one writes down and solves the Schrödinger equation for such a system, finding a nonzero probability that the electron may be found on the other side of the barrier and thus (one might suppose) "explaining" barrier penetration (or at least its possibility), this does not seem to involve tracing spatiotemporally continuous causal processes. Again, what does the notion of assembling information about causal processes and interactions amount to when one is dealing with systems with many interacting parts, such as a gas containing 10^{23} molecules or complex systems of the sort studied in neurobiology or economics? Whatever understanding the behavior of such systems involves, it must consist in more than just recording information about individual episodes of energy/momentum transfer via spatiotemporal contact (between, e.g., individual molecules in the case of the gas); instead it must involve finding tractable modes of representation that abstract from such details and represent (the relevant aspects of) their aggregative or cumulative impact. Such models may provide mechanical explanations in some broad sense that involve showing how the behavior of aggregates depends on features of local interactions among their parts, but this will usually involve going well beyond the very specific constraints on mechanical explanation imposed by Salmon's theory. Providing an account of explanation that remains in the broad spirit of the CM model but applies to such complex systems should be an important item on the agenda of those who wish to extend Salmon's work.

10. UNIFICATIONIST MODELS

These models draw their inspiration from the idea that there is a close connection between the extent to which a theory is explanatory and the extent to which it provides a unified treatment of a range of superficially different phenomena. This is a very intuitively attractive idea—unification is often regarded as an important scientific achievement or goal of inquiry. (Consider Newton's unification of terrestrial and celestial mechanics, the unification of electricity and magnetism begun by James Clerk Maxwell and perfected in the special theory of relativity, the unification of the electromagnetic and weak forces by Steven Weinberg and Abdus Salam, and so on.) Of course in developing this idea in a more systematic way, much will turn on exactly how the notion of unification is cashed out. In philosophy of science one of the earliest influential formulations of this approach, that of Michael Friedman (1974), understood unification as a matter of deriving a wide range of different explananda from a much "smaller" set of independently acceptable assumptions. Friedman's proposal was shown to suffer from various technical difficulties by Philip Kitcher (1976), who went on to develop his own version of the unificationist account in a series of influential essays (see especially Kitcher 1989).

Kitcher's basic idea is that successful unification is a matter of repeatedly using the same small number of argument patterns (that is, abstract patterns that can be instantiated by different particular arguments) over and over again to derive a large number of different conclusions—the fewer the number of patterns required, the more stringent they are in the sense of the restrictions they impose on the particular arguments that instantiate them, and the larger the number of conclusions derivable via them, the more unified the associated explanation. Thus Kitcher's model resembles the DN model in taking explanation to be deductive in structure but adds the further constraint that we should choose the deductive systemization of our knowledge that is most unifying among competing systemizations—it is the derivations provided by this systemization that are the explanatory ones.

Kitcher further claims that this additional constraint about unification allows us to avoid the standard counterexamples to the DN model, such as those described in section 7 above. Very roughly, this is because the unexplanatory derivations associated with the counterexamples turn out to use argument patterns that are less unified than the competing argument patterns that license derivations corresponding to our usual explanatory judgments. For example, according to Kitcher, derivations of facts about causes (the length of a pendulum) from facts about effects (the period of the pendulum) involve argument patterns and systemizations that are less unified than derivations that proceed in the opposite direction, and for this reason the latter are explanatory and the former are not.

More generally, Kitcher (1989, 477) claims that "the 'because' of causation is always derivative from the 'because' of explanation." That is, unification is what is primary; the causal judgment that, for example, the length of a pendulum causes its period is simply a consequence or reflection of our efforts at unification. There are no independent facts about causal relationships in nature to which our efforts at explanatory unification must be adequate. In this respect Kitcher returns to the original DN idea that the notion of explanation is more fundamental than the notion of causation.

Despite the intuitive appeal of the idea that there is a close connection between explanation and unification, providing a characterization of unification that captures our intuitive explanatory judgments has turned out to be far from straightforward. Part of the problem is that there are many different possible kinds of unification and only some of these seem to be connected to explanation—that is, there are nonexplanatory as well as explanatory unifications.[9] For example, one sort of "unification" consists in the use of the same mathematical structures and techniques to represent very different physical phenomena, as when both mechanical systems and electrical circuits are represented by means of Hamilton's or Lagrange's equations. This unified representation allows for the derivation of the behavior of both kinds of systems, but it would not be regarded by physicists as giving a common unified explanation of both kinds of systems or as an explanatory unification of mechanics and electromagnetism. Exactly what the latter involves is not entirely clear, but arguably it requires a demonstration that the same physical mechanisms or principles (perhaps more specifically the same forces) are at work in producing both

---

9. For a systematic development of this point and a far more detailed exploration of the relationship between explanation and unification, see Morrison 2000.

kinds of phenomena. (This after all is what is achieved by the Newtonian unification of terrestrial and celestial gravitational phenomena and by the unification of electricity and magnetism.) A closely related observation is that the Friedman/Kitcher conception of unification seems at bottom to be simply a notion of data compression or economical description—of finding a characterization of a set of phenomena or observations that allows one to derive features of them from a minimal number of assumptions. Many systems of classification (of biological species, diseases, personality types, etc.) seem to accomplish this without providing what seems intuitively to be explanations. Or to put the matter more cautiously, if they do provide explanations in some sense, these have a very different feel from the sorts of unifications provided by Newton and Maxwell.

A closely related observation, developed by several authors, is that it simply does not seem to be true that considerations of comparative unification always yield familiar judgments about causal asymmetries and causal irrelevancies—these seem to have (at least in part) an independent source.[10] Consider a theory, such as Newtonian mechanics, that is deterministic in both the past to future and the future to past directions and that contains time-symmetrical laws. It is far from obvious that derivations in such a theory that run from the future to the past—for example, from information about the future positions of the planets to their positions in the past—are any more or less unified than derivations that run from the past to the future, even though our intuitive judgments about preferred explanatory direction favor the latter explanations (cf. Barnes 1992). Considerations such as these suggest a possible alternative view about the role of unification in explanation: judgments about causal or explanatory dependence and about whether the same sort of dependence relationship is at work in different situations have some other source besides our efforts at unification. However, once we make such judgments, we can then go on to ask about the extent to which different phenomena are the result of the operation of the same mechanisms or dependency relations—to the extent that this is so, we have a unified explanation. Something like this "bottom up" picture of the role of unification in explanation is suggested

---

10. For details, see Barnes 1992 and Woodward 2003.

by Salmon (1989) and seems to fit at least some historical examples of successful unification.[11]

Even with this picture, however, there are puzzles.[12] Biological researchers originally worked out the mechanism of long-term potentiation or LTP (thought to be centrally involved in many forms of learning) in the sea snail, Aplysia. It was then found that the same mechanism underlies learning in many other biological species, including humans. Is this an example of explanatory unification? It certainly seems to conform to Kitcher's intuitive picture of unification—different phenomena (aspects of learning in many different biological species) are shown to result from the operation of the same fundamental mechanism.

On the other hand, how exactly should we understand the explanatory advance (if that was what it was) that was achieved? There is a natural thought that how well or badly the original account explains LTP in Aplysia should not depend on whether or not the same account also applies to other species. (Finding that this explanans applies to other species does not after all alter its content in any way. And if, contrary to actual fact, we were to find that the account applied only to Aplysia, with the explanation of long-term potentiation in other species taking a rather different form, why does this show that the account is any less good as an explanation of LTP in Aplysia.) Of course when we find that the account applies more widely, there is an obvious sense in which there is an explanatory advance—we now know the explanation of LTP in other species as well. But this seems to be a matter of being in a position to explain new explananda rather than (as Kitcher's discussion seems to suggest) having a better explanation of our original explanandum.[13] We also learn about new connections and relationships that were not previously understood, but

---

11. For an argument that this is the case for Newton, see Ducheyne 2005.
12. For related observations and additional discussion of some of the points that follow, see Woodward 2003 and Strevens 2007.
13. Suppose the explanatory improvement that takes place under unification consists just in our being able to explain new phenomena. Someone who is a skeptic about the explanatory virtue of unification could then argue that a theory that is not unified with our original theory could equally well explain these new phenomena. The moral seems to be that for Kitcher's story to work, the virtue of a more unified theory cannot be just that it enables us to explain things that were not previously explained. It must be that the unified theory explains things better than the disunified competitor.

again it is not entirely clear that this is an additional *explanatory* achievement rather than something else.

In this last connection it is worth noting that often one advantage or virtue of a relatively unified theory over less unified rivals is evidential: when a theory explains a range of different phenomena in terms of a single (or small set) of mechanisms or principles, there is, other things being equal, less room for "overfitting" in comparison with a theory that is allowed to resort to many disparate mechanisms. The requirement that a theory explain many different phenomena by reference to the same assumptions allows us to use those phenomena to sharply constrain the features of the postulated explanans and to bring evidence from many different sources to bear on it. This advantage is lost if each phenomenon is explained in a different way by reference to different assumptions. This observation prompts the thought that while it indeed may be a virtue or a good feature in a theory that it is unified, the virtue in question may have more to do with evidential support than with explanation.[14] On the other hand, the explanatory and evidential virtues may not be as sharply separable as this last remark assumes, especially given the rather elastic character of the notion of explanation. My rather inconclusive assessment is thus that although it is very plausible that explanation and unification are interconnected in important ways, more work needs to be done to characterize the kind of unification that is important for explanation and to spell out just what it is that unification gives us.

11. CONCLUSION

I conclude with some more general remarks about the current status of the topic of scientific explanation and possible directions for future work.

11.1. The Fate of the DN Model

The counterexamples to the DN model surveyed in section 7. are decades old and very widely known. Nonetheless, this model has shown

---

14. See Sober 2002 for an argument to this effect.

remarkable staying power. Although I know of no systematic surveys on the matter, my impression is that many philosophers (particularly in philosophy of physics and in areas of philosophy outside philosophy of science, including metaphysics) continue to believe that there must be something broadly right about the underlying idea of the DN model, even if the details may need to be tweaked in various ways. By contrast, philosophers of science who focus primarily on scientific disciplines other than physics, such as biology and the social sciences, tend to be more skeptical of the DN model.

Why is this? I conjecture that several factors are at work. First, it seems undeniable that in physics many paradigmatic textbook examples (examples that it seems natural to regard as cases of explanation) involve writing down a set of equations taken to describe some system of interest (where these equations are plausibly regarded as describing laws of nature), making assumptions about initial and boundary conditions, and then exhibiting a solution to the equations that corresponds to some behavior of the system one wants to understand. Illustrations include derivations of expressions for the fields set up by charged conductors of various shapes from the laws of classical electromagnetism and then derivations of the motions of charged particles within these fields; the various solutions to the field equations of general relativity (that is, derivations of features of the metrical structure of space-time from information about the mass-energy distribution within some region of space, various choices of boundary conditions, and the field equations themselves), such as the Schwarzschild solution, the Kerr solution, and so on; and the use of the Schrödinger equation to model various prototypical examples in elementary quantum mechanics (charged particle in a potential well, charged particle penetrating a potential barrier, and so on).

Each of these exercises seems to involve deriving a description of the behavior we are trying to understand from premises that include laws of nature—that is, it looks as though they use just the ingredients that figure in a DN explanation. Of course it is a jump to move from this observation to the conclusion that the DN model is the full story about *why* we find such derivations explanatory—that is, that such explanations work just by showing that their explananda are nomologically expectable and that nothing else about these derivations contributes to their explanatory import. Nonetheless, the ubiquity of deductive structures involving laws as premises contributes to many people's sense that the DN model corresponds to something real in scientific practice, at least in physics and perhaps in other areas of science where similar structures seem to play a role. By contrast, mathematical derivations from highly general principles seem to play a far less central role in disciplines like molecular biology and neurobiology (at least in their present stage of development). It is thus not surprising that the DN model seems less obviously applicable in such disciplines.

11.2. The Role of Causation

A second set of considerations bearing on the status of the DN model and the assessment of the various competing accounts of explanation has to do with the role of causal considerations in explanation. As we have seen, many of the standard counterexamples to the DN model seem to show that the model fails to incorporate certain commonsense causal distinctions (having to do with, e.g., the direction of causation). It is important to emphasize that this is *not* just a matter of the DN model failing to generate judgments that are in accord with our so-called intuitions. In many areas of inquiry (biomedicine, psychology, economics) the correct identification of causal direction and of relations of causal relevance is a central methodological concern. Assuming, for example, that there is a correlation between changes in the money supply M and economic output O, it is very important from the point of view of economics whether this correlation arises (i) because M causes O, (ii) because O causes M, or (iii) for some other reason. Among other considerations, this matters because these different claims about causal direction have very different implications, manipulation, and control. If (i) is correct, the Federal Reserve should be able, at least in principle, to manipulate O by intervening with the money supply; not so if (ii) is the correct analysis. From the perspective of a discipline like economics, it is hard to take seriously the idea that the difference between (i) and (ii) is somehow unreal or unimportant. So an account of explanation that is insensitive to the difference between (i) and (ii) will seem inadequate to economists and philosophers of economics and similarly for researchers in such other disciplines as biology and psychology. This is another reason philosophers who focus primarily on these disciplines often find it most natural to think of explanation as having to do with the identification of causal relationships and causal mechanisms rather than with the instantiation of DN-like structures.

It is arguable that these issues about the role of causal considerations in explanation look rather different if one takes the paradigmatic science to be fundamental physics. Although the topic deserves a far more detailed treatment than I can give it here, there is a widespread (although by no means universal) belief among philosophers of physics that a rich and thick notion of causation of the sort found in common sense (or at least important features of this notion) fail to apply in a straightforward or unproblematic way in certain fundamental physics contexts or at least that this causal notion is not "grounded" simply in facts about fundamental physical laws. (For various views of this character although they differ among themselves in important ways see, e.g., Russell 1913; Field 2003; Norton 2007, and for additional discussion, see Woodward 2007.) I take no stand here on whether this belief is correct but merely observe that to the extent that one regards it as correct, the failure of the DN model to fully incorporate commonsense causal distinctions may look like a virtue of that model or at least not a serious defect.

One of the clearest illustrations of this point is provided by the asymmetry of the causal relationship. I have suggested that in disciplines like biology and economics causal claims that reflect this asymmetry are pervasive and, to the extent that they can be tied to asymmetries of manipulation and control, are often taken to be unproblematic. Both the ultimate physical origin(s) of such asymmetries (and whether indeed there is a unified story to be told about their origin) and the possibility that their range of application may be limited in some way are not issues that fall within the purview of these disciplines. By contrast, issues of this sort do of course arise in fundamental physics and seem to have implications for how we should think about fundamental physical explanation and the role of causal considerations in such explanation. To take only one of the most obvious possibilities, suppose that the causal asymmetries have their origin in thermodynamic asymmetries in some way (cf.

Albert 2000; Kutach 2007). Then one might well wonder whether causal asymmetries have any application to sufficiently simple microscopic systems governed by time-symmetrical laws that lack a well-defined direction of entropy increase. Perhaps for such systems there are symmetrical relations of dependence, grounded in physical law, but no objective basis for picking out one event as asymmetrically related to another as cause to effect. If so, a model of explanation that applies to such systems presumably should not incorporate directional features. My point is *not* that we should regard this last claim as uncontroversially correct but rather that *if* one takes it to be correct (or at least takes it to be a serious possibility), then the failure of a model of physical explanation to incorporate commonsense considerations about causal asymmetries need not automatically be regarded as a fundamental defect. One might well take the view that until we have a better understanding of the physical basis and significance of the causal asymmetries we see in everyday life (and relatedly the range of systems in which the appropriate physical basis for the asymmetries is present), it is premature to insist that these asymmetries must be incorporated into a model of explanation that is applicable to fundamental physics. Similarly for other features of commonsense causal reasoning. Thus one development that would contribute to further progress in connection with the topic of scientific explanation is a better understanding of the legitimate role (if any) that causal considerations play in fundamental physics. Relatedly, those who are skeptical about the role of causal considerations in fundamental physics should spell out the implications of their views for physical explanation.

11.3. Laws Revisited

We noted above that there is considerable disagreement about the criteria that distinguish laws from other true generalizations and about the role of laws in the various scientific disciplines. This in turn raises some obvious and still unresolved questions for the DN and other law-based models of explanation. Putting aside the view that the whole notion of a law of nature rests on a confusion, it might appear that in physics at least there often will be agreement in particular cases about whether generalizations count as laws, even if we lack an adequate general theory of lawfulness: the Schrödinger equation and the Klein-Gordon equation count as laws if anything does, and Bode's "law" does not. However, in the absence of a more adequate general theory, a nontrivial difficulty arises when one tries to apply the DN and other law-based models to sciences like biology, psychology, and economics. Virtually everyone agrees that these disciplines contain true causal generalizations that can figure in explanations, but there is a great deal of controversy about whether these generalizations should be regarded as laws. For example, the "laws" of segregation and independent assortment in Mendelian genetics have well-known exceptions and appear to be contingent on the course of evolution in the sense that these generalizations would not apply to biological organisms even to the extent that they do had those organisms been subject to sufficiently different selective pressures (cf. Beatty 1997). Does this mean that we should not think of these generalizations as laws? Philosophers of biology have been sharply divided on this question, and one finds a parallel debate concerning the status of the generalizations that figure in the other special sciences.[15]

Taken literally and strictly, the DN and other nomothetic models seem to imply that scientific disciplines that do not contain laws do not provide explanations. If this conclusion is unacceptable, then there seem to be two remaining possibilities: (i) either there are laws in the special sciences after all (and we need to provide an account that identifies such laws and explains how to distinguish them from nonlaws), or else (ii) we need to formulate an alternative to law-based models of explanation that allows the special sciences to furnish explanations and causal knowledge even though they do not contain laws. With respect to (i), the challenge, in my opinion, is to provide an account that both (a) includes in the category of laws those generalizations that are plausibly regarded as explanatory and (b) at the same time does not include nonexplanatory generalizations. Most current proposals seem to fail along dimension (b).[16]

15. For biology, see, for example, Beatty 1997; Brandon 1997; Sober 1997; Mitchell 2000. For the social sciences, see Kincaid 1989.
16. For example, it is arguable that this is true of current attempts to show that the generalizations of the special sciences are "ceteris paribus" laws. See Earman et al. 2002 and Woodward 2002.

I noted above that to the extent that we lack a compelling account of what laws are we will also lack insight into whether (and, if so, why) laws are required for explanation. In this connection it is worth considering the account of laws that is currently regarded as the most promising by many philosophers, the MRL theory. According to this theory, laws are those generalizations that figure as axioms or theorems in the deductive systemization of our empirical knowledge that achieves the best combination of simplicity and strength (where strength has to do with the range of empirical truths that are deducible).[17] If we ask about the model of explanation with which this treatment of laws most naturally fits, the answer seems to be unificationist approaches: if laws are generalizations that play a central role in the achievement of simple (and presumably unified) deductive systemizations, then by appealing to laws in explanation we achieve explanatory unification—this makes it intelligible why it is desirable that explanations invoke laws.[18] If an account along these lines could be made to work (that is, if we had a good defense of the MRL theory *and* of the idea that explanation involves unification), then we would have the sort of integrated story about laws and explanation that I claimed was largely lacking in the DN account—a story about what laws are that is directly connected to a story about the point of explanation. I do not claim that the story just sketched is unproblematic—we have already noted that there are difficulties with the unificationist model, and there are also problems with the MRL theory of laws as well[19]—but it does represent an exemplar of the sort of integrated story that advocates of law-based models should be attempting to tell. In this connection it is also worth asking what the implications of the MRL account are for the lawfulness of the generalizations of the special sciences. In my view, it is far from obvious that the MRL account yields

---

17. For a statement of the basic idea, see Lewis 1973.
18. Notice, though, that although this rationale yields an account of why it is *desirable* to construct explanations appealing to laws (when this is possible), it is not clear that it yields the result that explanation always *requires* laws. Perhaps there are generalizations that unify sufficiently to qualify as explanatory, according to the unificationist account, but do not count as laws according to the MRL account. In this connection it is interesting that Kitcher's version of the unficationist model explicitly does not require laws.
19. See Woodward 2003, 288–295, 358–373.

the result that laws are common in the special sciences, but it would be very useful to have a more systematic exploration of this issue.

References

Albert, D. 2000. *Time and Chance*. Cambridge, MA: Harvard University Press.
Barnes, E. 1992. "Explanatory Unification and the Problem of Asymmetry." *Philosophy of Science* 59 (4): 558–571.
Beatty, J. 1997. "Why Do Biologists Argue like They Do?" *Philosophy of Science* 64 (supp.): S432–S443.
Bechtel, W., and A. Abrahamsen. 2005. "Explanation: A Mechanistic Alternative." *Studies in the History and Philosophy of Biological and Biomedical Sciences* 36 (2): 421–441.
Brandon, R. 1997. "Does Biology Have Laws?" *Philosophy of Science* 64 (supp.): S444–S457.
Bromberger, S. 1966. "Why Questions." In *Mind and Cosmos: Essays in Contemporary Science and Philosophy*, ed. R. Colodny, 86–111. Pittsburgh: University of Pittsburgh Press.
Cartwright, N. 1983. *How the Laws of Physics Lie*. Oxford: Clarendon.
Cartwright, N. 2004. "From Causation to Explanation and Back." In *The Future for Philosophy*, ed. B. Leiter, 230–245. Oxford: Oxford University Press.
Dowe, P. 2000. *Physical Causation*. Cambridge: Cambridge University Press.
Ducheyne, Steffen. 2005. "Newton's Notion and Practice of Unification." *Studies in History and Philosophy of Science*, pt. A, 36 (1): 61–78.
Earman, J. 1993. "In Defense of Laws: Reflections on Bas van Fraassen's *Laws and Symmetry*." *Philosophy and Phenomenological Research* 53 (2): 413–419.
Earman, J., J. Roberts, and S. Smith. 2002. "Ceteris Paribus Lost." *Erkenntnis* 57 (3): 281–301.
Field, H. 2003. "Causation in a Physical World." In *The Oxford Handbook of Metaphysics*, ed. M. Loux and D. Zimmerman, 435–460. Oxford: Oxford University Press.
Friedman, M. 1974. "Explanation and Scientific Understanding." *Journal of Philosophy* 71 (1): 5–19.
Giere, R. 1999. *Science without Laws*. Chicago: University of Chicago Press.
Hausman, D. 2002. "*Physical Causation*." *Studies in History and Philosophy of Modern Physics*, pt. B, 33 (4): 717–724.
Hempel, C. 1965a. "Aspects of Scientific Explanation." In *Aspects of Scientific Explanation and Other Essays in the Philosophy of Science*, 331–496. New York: Free Press.
Hempel, C. 1965b. *Aspects of Scientific Explanation and Other Essays in the Philosophy of Science*. New York: Free Press.

Hempel, C. 1965c. "The Logic of Functional Analysis." In *Aspects of Scientific Explanation and Other Essays in the Philosophy of Science*, 297–330. New York: Free Press.

Hitchcock, C. 1995. "Discussion: Salmon on Explanatory Relevance." *Philosophy of Science* 62 (2): 304–320.

Kincaid, H. 1989. "Confirmation, Complexity, and Social Laws." In *PSA 1988*, ed. A. Fine and J. Leplin, 299–307. East Lansing, MI: Philosophy of Science Association.

Kitcher, P. 1976. "Explanation, Conjunction, and Unification." *Journal of Philosophy* 73 (8): 207–212.

Kitcher, P. 1989. "Explanatory Unification and the Causal Structure of the World." In *Scientific Explanation*, ed. P. Kitcher and W. Salmon, 410–505. Minnesota Studies in the Philosophy of Science 13. Minneapolis: University of Minnesota Press.

Kutach, D. 2007. "The Physical Foundations of Causation." In *Causation, Physics, and the Constitution of Reality: Russell's Republic Revisited*, ed. H. Price and R. Corry, 327–350. Oxford: Oxford University Press.

Lewis, D. 1973. *Counterfactuals*. Cambridge, MA: Harvard University Press.

Machamer, P., L. Darden, and C. Craver. 2000 "Thinking about Mechanisms." *Philosophy of Science* 67 (1): 1–25.

Mitchell, S. 2000. "Dimensions of Scientific Law." *Philosophy of Science* 67 (2): 242–265.

Morrison, M. 2000. *Unifying Scientific Theories*. Cambridge: Cambridge University Press.

Norton, J. 2007. "Causation as Folk Science." In *Causation, Physics, and the Constitution of Reality: Russell's Republic Revisited*, ed. H. Price and R. Corry, 11–44. Oxford: Oxford University Press.

Russell, B. 1913. "On the Notion of Cause." *Proceedings of the Aristotelian Society* 13:1–26.

Salmon, W. 1971. "Statistical Explanation." In *Statistical Explanation and Statistical Relevance*, ed. W. Salmon, 29–87. Pittsburgh: University of Pittsburgh Press.

Salmon, W. 1984. *Scientific Explanation and the Causal Structure of the World*. Princeton, NJ: Princeton University Press.

Salmon, W. 1989. *Four Decades of Scientific Explanation*. Minneapolis: University of Minnesota Press.

Salmon, W. 1994. "Causality without Counterfactuals." *Philosophy of Science* 61 (2): 297–312.

Salmon, W. 1997. "Causality and Explanation: A Reply to Two Critiques." *Philosophy of Science* 64 (3): 461–477.

Scriven, M. 1959. "Explanation and Prediction in Evolutionary Theory." *Science* 130 (3374): 477–482.

Scriven, M. 1962. "Explanations, Predictions, and Laws." In *Scientific Explanation, Space, and Time*, ed. H. Feigl and G. Maxwell, 170–230. Minnesota Studies in the Philosophy of Science 3. Minneapolis: University of Minnesota Press.

Sklar, L. 1999. "The Content of Science, the Methodology of Science, and Hempel's Models of Explanation and Confirmation." *Philosophical Studies* 94 (1–2): 21–34.

Sober, E. 1997. "Two Outbreaks of Lawlessness in Recent Philosophy of Biology." *Philosophy of Science* 64 (supp.): S458–S467.

Sober, E. 2002. "Two Uses of Unification." In *The Vienna Circle and Logical Empiricism*, ed. F. Stadler, 205–216. Vienna Circle Institute Yearbook. Dordrecht, Netherlands: Kluwer Academic.

Spirtes, P., C. Glymour, and R. Scheines. 2000. *Causation, Prediction, and Search*. 2nd ed. Cambridge, MA: MIT Press.

Strevens, M. 2007. "An Argument against the Unification Account of Explanation." Draft of December, 2009. www.strevens.org/research/expln/Unificatio.pdf. Accessed April, 2014.

van Fraassen, B. 1989. *Laws and Symmetry*. Oxford: Clarendon.

Woodward, J. (2002) "There's No Such Thing as A Ceteris Paribus Law." *Erkenntnis* 57 (3): 303–328.

Woodward, J. 2003. *Making Things Happen: A Theory of Causal Explanation*. New York: Oxford University Press.

Woodward, J. 2007. "Causation with a Human Face." In *Causation, Physics, and the Constitution of Reality: Russell's Republic Revisited*, ed. H. Price and R. Corry, 66–105. Oxford: Oxford University Press.

Chapter 2

Probabilistic Explanation

MICHAEL STREVENS

1. VARIETIES OF PROBABILISTIC EXPLANATION

Science turns to probabilistic, as opposed to deterministic, explanation for three reasons. Most obviously, the process that produces the phenomenon to be explained may be irreducibly indeterministic, in which case no deterministic explanation of the phenomenon will be possible, even in principle. If, for example, the laws of quantum mechanics are both probabilistic and fundamental—as most scientists believe—then any explanation of, say, an episode of radioactive decay can at best cite a very high probability for the event (there being a minuscule probability that no atom will ever decay). The decay, then, must be explained probabilistically.

Because all the world's constituents conform to quantum dictates, it might seem that, for the very same reason, everything must be given a probabilistic explanation. For many phenomena involving large numbers of particles, however, the relevant probabilities tend to be so close to zero and one that the processes producing the phenomena take on a deterministic aspect. It is traditional in the philosophy of explanation to treat the corresponding explanations as deterministic.

Thus, you might think, there will be a simple division of labor between probabilistic and deterministic explanation: probabilistic explanation for phenomena involving or depending on the behavior of only a few fundamental-level particles, due to the indeterministic aspect of quantum mechanical laws, and deterministic explanation for higher-level phenomena where quantum probabilities effectively disappear. However, even high-level phenomena are routinely given probabilistic explanations—for reasons that have nothing to do with metaphysical fundamentals.

In some cases, the recourse to probability is for epistemic rather than metaphysical reasons. Although the phenomenon to be explained is produced in an effectively deterministic way, science's best model of the process may be missing some pieces and so may not predict the phenomenon for sure. In such a case, the explanation is typically given a probabilistic form. Whatever the model says about the phenomenon is put in statistical terms—perhaps as a probability of the phenomenon's occurrence or as a change in the probability of the phenomenon's occurrence brought about by certain factors—and these statistical facts are offered as a partial explanation of what has occurred.

There are many examples to be found in medicine. If a heavy smoker contracts emphysema, his or her smoking is typically cited as a part of the explanation of the disease. Smoking probabilifies emphysema, but we do not know enough about its etiology to see for sure whether any particular heavy smoker will become emphysemic. Thus our best explanation of a heavy smoker's emphysema must be probabilistic. Though we will perhaps one day be able to do better, we find the present-day probabilistic explanation enlightening: if it is not the best possible explanation of emphysema, it is certainly a fairly good explanation.

A third occasion for probabilistic explanation arises in certain cases where the process producing the phenomenon to be explained is rather complex and could have produced that very phenomenon in a number of different ways. In such cases, there appears to be explanatory value in a description of the process that abstracts away from the details that determine that the phenomenon occurred in the particular way that it did and that presents only predisposing factors that make it highly likely that some such process would occur.

Perhaps the best examples are to be found in statistical physics. To explain why a gas rushes into a vacuum so as to equalize its density everywhere, you might recount the deterministic details in virtue of which each particle ends up in a state that, in the large, constitutes an equalization of density. You would have to cite the initial position and velocity of each particle and derive its later position from these initial conditions using the appropriate laws of molecular dynamics.

Statistical physics tells the story in a different way. It can be shown (though the details are disputed) that almost every set of initial conditions leads a gas to equalize its density. The demonstration invokes only some quite general properties of molecular dynamics, so it is far simpler than the demonstration suggested in the previous paragraph. But because the *almost every* is not an *every*, it is not deterministic; statistical physics, as the name suggests, gives the explanation an explicitly probabilistic cast.

If the statistical explanation of density equalization were less satisfactory than the deterministic explanation, this case could be assimilated to the emphysema case as a use of probability to fill in, for the time being, a gap left by scientific ignorance. But most writers (though certainly not all) would agree that the statistical explanation is superior to the deterministic explanation: you do not fully understand why gases fill vacuums until you see why almost any set of initial conditions leads to equalization, and once you appreciate this fact, seeing that the particular initial conditions of your particular gas led to equalization adds little or nothing of explanatory value. Here, then, probability is introduced into an explanation not because it is either metaphysically or epistemologically unavoidable but because it enhances the explanation, providing more insight than the deterministic alternative. Even if we were able to construct a deterministic explanation of some gas's expansion, we would prefer the statistical explanation.

Another sort of explanation that perhaps belongs in the same class as the explanations of statistical physics is the probabilistic explanation of frequencies of outcomes obtained on simple gambling devices. To explain the fact that a large number of coin tosses, for example, turns up heads about one-half of the time, we typically cite the one-half probability of a tossed coin landing heads, though we know that coin tossing is an effectively deterministic process. From the one-half probability we can derive a very high probability of a frequency of heads approximately equal to one-half but without being able to make any predictions about the particular sequence of heads and tails that realizes the frequency. As in the case of the gas, we extract from our knowledge of the physics of coin tosses the fact that almost any set of initial conditions for a long series of tosses will produce a frequency of heads of about one-half (Strevens 1998), and we prefer the corresponding explanation to a deterministic explanation that begins with the exact initial conditions of the series, derives a particular sequence of heads and tails, and calculates the frequency of heads in the sequence.

A philosophical account of probabilistic explanation must do some kind of justice to the three varieties of probabilistic explanation I have described. It must show why probabilistic explanation is acceptable if the world is indeterministic; it must show why it is also acceptable in cases where our knowledge of the world precludes our constructing a deterministic explanatory model; and it must show why probabilistic explanation is preferable to deterministic explanation in statistical physics and certain other domains.

Then again, an account of probabilistic explanation might be in part a debunking of some or all of these claims, for example, an argument that scientists and philosophers are mistaken in thinking that probabilistic explanation is ever preferable to deterministic explanation—or even an argument that there can be no such thing as probabilistic explanation. Deflationary arguments of this sort can be found in the literature, but most theories of probabilistic explanation attempt to make sense of, rather than to deny the existence of, successful probabilistic explanation in quantum mechanics, medical science, and statistical physics.

Accounts of scientific explanation can be classified in two ways: according to their conception of the nature of explanation or according to the formal criteria they impose on an explanation. These properties are of course linked but not especially strongly. Two theorists might agree that an explanation should consist of a list of factors that are statistically relevant to the phenomenon to be explained (a formal criterion) yet disagree on the question why a statistically relevant factor casts light on the phenomenon. Or they might agree that explanation is ultimately concerned with giving the causes of a phenomenon yet disagree on the question whether the causes should be presented in the form of a logical argument.

A treatment of approaches to explanation that follows the traditional divisions in the literature results, for this reason, in a certain amount of cross classification: as you will see, some accounts of explanation might reasonably be placed under more than one of the organizational headings employed in this essay. When Carl G. Hempel initiated the systematic study of probabilistic explanation in 1963, however, he clearly specified both a formal criterion for probabilistic explanation and a doctrine of the nature of the underlying explanatory relation. This provides a place to begin the survey of approaches to probabilistic explanation that is both especially firm and chronologically apt.

2. NOMIC EXPECTABILITY AND THE INDUCTIVE-STATISTICAL ACCOUNT

An explanation, Hempel (1965, 337) writes, "shows that, given the particular circumstances and the laws in question, the occurrence of the phenomenon *was to be expected*; and it is in this sense that the explanation enables us to *understand why* the phenomenon occurred." Although this passage concerns deterministic explanation, Hempel's (1965, sec. 3) treatment of probabilistic explanation is also based on the posit that scientific understanding is a matter of nomically based expectation: a probabilistic explanation, Hempel proposes, uses a statistical rather than a deterministic law to show that the phenomenon to be explained was to be expected.

From this conception of nomic expectability as the underlying explanatory relation, Hempel extracts a formal criterion for probabilistic explanation. Or rather he extracts two criteria, one for the explanation of singular events and one for the explanation of laws. A probabilistic law can be explained by deriving the law deductively from other laws and (if necessary) initial conditions. This is precisely the same formal criterion that Hempel gives (tentatively—see Hempel and Oppenheim 1948, 159, n28) for the explanation of a deterministic law.

PROBABILISTIC EXPLANATION

The formal criterion for the probabilistic explanation of singular events constitutes, by contrast, a significant departure from the deterministic case. For this reason and because the probabilistic explanation of singular events perhaps raises more questions than the probabilistic explanation of laws, philosophers writing on probabilistic explanation have focused on the explanation of events. This essay will of necessity reflect the bias toward event explanation.

An event explanation must show, according to Hempel's doctrine of nomic expectability, that the event to be explained was to have been expected. As a consequence, an explanation functions like, and for Hempel is identical to, an argument. If the event is produced deterministically, the argument can be deductive. If not, it can only be (if not trivial) inductive. Thus Hempel proposes that a probabilistic explanation of an event is a sound law-involving inductive argument to the effect that the event occurred. This he calls the inductive-statistical or IS account of probabilistic event explanation. The criteria for the soundness of an inductive argument differ in interesting ways from the criteria for the soundness of a deductive argument; many of the more controversial features of the IS account arise from the differences.

Suppose (Hempel's example) that you are assigned to explain why Jen Jones, who contracted a streptococcus infection, recovered within a week. It turns out that she was given penicillin and that 90 percent of all strep patients recover quickly, that is, within a week, when given penicillin. You explain Jones's swift recovery probabilistically by citing its high probability, given the administration of penicillin.

According to Hempel, your explanation should be understood as an inductive argument with the event to be explained—that Jen Jones recovered within a week—as its conclusion and the relevant probabilistic facts as its premises. Hempel formalizes the argument as follows:

Jen Jones had a strep infection
Jen Jones was given penicillin
90 percent of all strep patients recover within a week when treated with penicillin

---

Jen Jones recovered within a week

Because the argument is not deductive, the premises do not entail the conclusion; rather they bestow a 90 percent inductive probability on the conclusion. This is high enough, as Hempel writes, that you should have expected Jones to recover. Because your expectation is based in part on a statistical law, it confers scientific understanding.

Hempel's probabilistic event explanations differ from his deterministic event explanations in two ways, corresponding to two ways inductive logic differs from deductive logic. First is the matter of the probability itself. An inductive argument must confer a sufficiently high probability on its conclusion to justify an expectation that the conclusion holds. Typically, it is required that the probability be greater than 50 percent. It follows, on the IS account, that events with a probability of 50 percent or less cannot be explained. Since the probability of even a heavy smoker contracting emphysema is less than 50 percent, for example, it seems that citing smoking alone cannot explain emphysema, though citing smoking together with other predisposing factors might be sufficiently probabilifying to be explanatory. That is a controversial claim.

Second, it is not enough, in order to provide a sound inductive argument that the recovery will occur, to find some collection of facts that together probabilify the recovery to a degree greater than 50 percent. The facts must in a certain sense be complete. To illustrate this point, Hempel asks you to imagine that Jones's infection is known to be resistant to penicillin. Suppose that the probability of swift recovery when infected with a penicillin-resistant strain is 10 percent whether or not penicillin is administered. Then you ought not to have expected Jones to recover swiftly.

The fact of penicillin resistance in no way undermines the truth of the premises of the inductive argument spelled out above. It does undermine the argument itself; typically it is said that an inductive argument is not sound unless it includes all relevant information in its premises. Because the argument for swift recovery omits the information concerning penicillin resistance, it does not provide inductive support for its conclusions. Thus the power of an inductive argument depends not only on its internal structure and on the truth of its premises but also on the stock of background knowledge.

For essentially the same reasons, Hempel claims that the argument does not, in the penicillin resistance case, explain its conclusion, a consequence reflected in actual explanatory practice. It is not possible for Hempel to require that an IS explanation cite all relevant background knowledge among its premises, since the fact that the explanandum occurred is normally known and so would have to be included, making the argument deductive and, because not law involving, unexplanatory. Other related items of inductively relevant information must also be excluded as a matter of course from the explanatory argument, for example, evidence that the explanandum occurred. Hempel formulates what he calls the *requirement of maximal specificity* to find the right balance of inclusion and omission. The details will not be discussed here.

The requirement of maximal specificity demands that (almost) all relevant background knowledge be included in an inductive argument if it is to serve an explanatory function, but unknown facts that would be inductively relevant if known do not fall within the scope of the requirement. Thus if Jen Jones's infection is resistant to penicillin but this fact is not known to the medical community, the argument above qualifies as a good explanation. Certainly it is reasonable to count the argument as a piece of sound inductive reasoning—inductive reasoning is always provisional—but it does not seem so reasonable to say that penicillin explains Jones's recovery provided that her doctors do not discover that hers is a resistant strain. The doctors may think that they are giving a good explanation, but they are wrong—so you would like to say. Hempel's account, however, has the consequence that they are genuinely giving a good explanation; only if they find out about the resistance does their explanation cease to confer understanding.

These, then, are the two principal objections to the IS account of event explanation (ignoring, for the sake of this essay, objections that apply equally to Hempel's deductive-nomological account of event explanation, such as the charge that an expectability account illegitimately allows effects to explain their causes): first, that the IS account allows only events with relatively high probabilities to be explained, whereas science considers itself able to explain low-probability events, and second, that the IS account relativizes probabilistic explanation to an epistemic background where we find in scientific practice no such relativity. The high-probability objection has spawned a debate that has been conducted to some extent independently of the rest of the philosophy of probabilistic event explanation, a debate that concerns the extent to which the factors cited in an explanatory argument must probabilify the event to be explained and whether factors that lower the probability ought to appear in an explanation (as on the IS account, you will note, they must). I defer substantive discussion of the debate to section 5. The epistemic relativity objection has motivated a search for a conception of the explanatory relation and a concomitant account of probabilistic explanation, according to which the probabilities involved in explanation are not inductive probabilities but something more objective, usually real physical probabilities.

3. CAUSAL APPROACHES

Famous counterexamples to Hempel's deductive-nomological account of explanation and more broadly to his conception of explanation as nomic expectability, such as the case of the length and period of a pendulum (see James Woodward's chapter "Scientific Explanation" in this volume), have by now convinced the great majority of philosophers that an account of explanation must provide a starring, if not exclusive, role for causation. The simplest way to do so is to hold that what explains an event are its causes and the background conditions and laws (or causal generalizations) in virtue of which they are causes.

The causal conception has naturally been transplanted to the territory of probabilistic event explanation, but it is not clear exactly where it should take root. Most commonly, it is suggested that physical probabilities (sometimes called *chances*) can be understood as a kind of causal disposition. A probability explains the outcome that it is the probability of, on this view, the same way that any disposition explains the event that it is a disposition to produce—in the same way, for example, that a magnet's disposition to attract any nearby iron filings explains a clump of filings jumping and clinging to the magnet. Among the dispositionalists may be counted Alberto Coffa (1974), James Fetzer (1974), and as godfather if not participant, Richard Jeffrey (1971).

Whereas in the IS model it is inductive probabilities that explain, in a dispositional theory it is physical probabilities, interpreted as a kind of cause. This gives the dispositionalist a more satisfying account of the case of penicillin resistance, discussed in the previous section, than Hempel is able to offer. If Jen Jones's infection is not known to be penicillin resistant, then her treatment with penicillin confers, uncontroversially, a high inductive probability on her swift recovery. The IS account is committed, controversially, to counting this probability as explanatory. Dispositionalists are not: they require that the penicillin, if it is to contribute to the explanation, should contribute to a physical disposition to recovery; in the case where the infection is resistant, it clearly does not.

A dispositional view of probability and probabilistic explanation can also avoid the worst consequences of the IS account's requirement that an event have a high probability if it is to be probabilistically explained in either of two ways. First, an explanation may be allowed to cite less than the full complement of causes (though presumably more causes are on the whole better). If you explain a case of emphysema by pointing to heavy smoking, you indicate one cause of the emphysema—one part of the basis for the probability of the smoker contracting, thus by hypothesis the smoker's disposition to contract, emphysema—and so give a genuine, if suboptimal, explanation of the emphysema. On the IS account, by contrast, your attempt fails to count as an explanation at all.

Second, in the case where an event's only cause fails to probabilify it highly, on a causal account the citation of that cause and thus of the low probability nevertheless constitutes a causally exhaustive hence an optimal explanation of the event. Suppose, for example, that the probability of a certain atom undergoing radioactive decay in a given time period is low. On the IS account, the event of the decay occurring within the time period, if it happens, cannot be explained. On a causal approach it can be given a complete explanation by delineating the relevant causal information, that is, the aspects of the structure of the atom and the laws of quantum mechanics that fix the probability of decay.

Many commentators have found the idea of a probabilistic disposition obscure. Whereas the familiar, deterministic dispositions, such as fragility and paramagnetism, can be given a counterfactual analysis, any parallel analysis of probabilistic dispositions seems itself to contain an ineliminable reference to probability.

It is not necessary to understand probabilities as dispositions in order to take a causal approach to probabilistic explanation. One alternative, suggested by Paul Humphreys, takes the factors that affect the value of an event's probability but not the probability itself as the causes of the event. Heavy smoking is a cause of emphysema, then, because it increases the probability of emphysema, but the probability itself is not a causal disposition. Indeed—taking an extremely deflationary attitude to probabilities—Humphreys (1989) declares them to be "literally nothing" (141).

A different approach, which is compatible with either Humphrey's deflationism or dispositionalism, is to hold that when probabilities enter into causal explanation, they do so in the guise of probabilistic causal laws. Heavy smoking is a cause of emphysema, on this view, because there is a causal law about emphysema that connects it probabilistically to smoking. Thus probabilities are a part of causal explanation, because the causes do their causing probabilistically; it is left open whether probabilities are themselves causal. To the philosopher of explanation, this view has the advantage of leaving some of the more contentious issues to the metaphysicians. In practice, however, it has not been so easy for explanatory causalists to avoid the metaphysics of probability (this writer included).

4. PROBABILISTIC RELEVANCE ACCOUNTS

What probabilistic relevance accounts have in common is not a conception of the explanatory relation—as expectability, causality, or something else—but an aspect of the formal criterion they offer for a good explanation. They are agreed on two things. First, they hold that a probabilistic explanation is either a list of factors that are probabilistically relevant to the event to be explained or some other structure designed to exhibit such factors and their relevance, such as a deductive argument. Second, they take this fact about the formal criterion to be a more fundamental and reliable datum about explanation than any intuitions you might have about the underlying nature of scientific understanding. They differ, then, from an account such as Humphreys's, which also requires that an explanation cite probabilistically relevant factors, by leaving open the question of the nature of the explanatory relation.

Hempel's formal criteria for both deterministic and probabilistic explanation contain no effective safeguard against the inclusion in an explanatory argument of irrelevant information, that is, in the case of event explanation, information that plays no role in entailing the occurrence of the event to be explained (in the deterministic case) or probabilifying its occurrence (in the probabilistic case). Its appearance in an explanatory argument implies that such information is explanatorily relevant, but intuition suggests otherwise. The expectability approach could be insulated against this objection in the following way: an explanation is a list of factors that, first, provide good reason to believe that the phenomenon to be explained occurs and, second, are statistically relevant to its occurrence.

Wesley Salmon suggests instead a radical departure from Hempel's argument-centered formal criterion for explanation: all that matters for an explanation is the satisfaction of the second requirement, relevance to the event's occurrence. Further, Salmon (1971) suggests, the relevance relation in question is not epistemic but physical: a factor is relevant to the occurrence of an event if it affects the physical probability of the event.

For explanatory purposes, a factor d is not considered to affect the probability of an event e if d is screened off from e by a further event c but not vice versa, meaning that d does not affect the probability of e in the presence of c but c does affect the probability of e in the presence of d. Thus, for example, d is not said to affect the probability of e in the case where c is a common cause of d and e, despite the fact that d and e will typically be correlated.

According to what Salmon calls the statistical relevance account, an event explanation is a list or table of all factors that make a difference to the physical probability of an event occurring. Salmon takes an expansive view of this table of relevance: he suggests that it should include information about factors that were relevant to unrealized alternatives to the event to be explained and factors that were not present but that would have been relevant if present. Factors that are negatively relevant are also considered explanatory; that is, factors that lower the probability of an event's occurrence should be included in the event's explanation. Salmon defends these requirements only in passing; they may be regarded as optional elements of the statistical relevance account.

Peter Railton (1978) has also offered what may be considered a statistical relevance account of probabilistic event explanation, though he calls his theory the deductive-nomological-probabilistic account. Railton proposes that an event is explained by deducing its physical probability from the relevant laws and initial conditions. In contrast to the IS account, the explanatory argument is deductive and has the physical probability of the event as its conclusion rather than the event itself. Further, the argument is explicitly required to contain only facts essential to the deduction. A Railtonian explanation, then, will contain only facts found in a Salmon explanation arranged in the form of a deduction. It will not contain all the facts that Salmon requires, however: it will mention only factors that were present and that contributed to the probability of the event to be explained (as opposed to its unrealized alternatives).

Salmon and Railton do not link their accounts to a particular conception of the explanatory relation. Unofficially, Salmon perhaps takes statistical relevance itself as the relation, at least in his early work, and Railton occasionally talks in a causal idiom. But officially their accounts are open to many interpretations, for example, to either dispositionalism or its denial.

Neither account, however, is compatible with Hempel's expectability conception of the explanatory relation, since both allow the probability of the explained event to be as low as you like. Provided that you can cite all the factors that play a part in determining the probability of an event, you can explain the event, even if it is very unlikely.

Peter Achinstein (1983) presents the following counterexample to any probabilistic relevance account of event explanation (including theories such as Humphreys's). Petra takes poison. This particular poison has a 90 percent chance of killing anyone who takes it within twenty-four hours. As it happens, Petra survives the poisoning but is run over by a bus exactly twenty-four hours after taking the poison. The poison probabilifies her death but does not explain it. (Stuart Gluck and Steven Gimbel [1997] offer a more sophisticated version of this argument.) Thus there is more to explanation than probabilistic relevance.

In reply a proponent of probabilistic relevance might argue that there must have been something about Petra—her high metabolism or her having breakfasted on the antidote—that prevented her dying. In the context of this intervening factor, the poison did not raise the probability of death after all, thus the probabilistic relevance approach does not count it as explanatorily relevant. Surely, however, the action of the poison might be genuinely indeterministic. Or it might be deterministic but of a piece with the processes that are the subject matter of statistical physics. Either way, you must conclude that the probability of death was raised but nothing came of it and that the probability raiser is as a consequence no part of the explanation of death. The same problem arises for accounts of probabilistic causation (Menzies 1996); for a solution, see Strevens 2008, section 11.3.

5. ELITISM AND EGALITARIANISM

Elitism is the blanket term I give to a preference for high probability in explanation; *egalitarianism* is a contrary indifference to probabilistic magnitude and, in the case of probability changes, perhaps even to sign. There are two questions in particular toward which elitist and egalitarian attitudes can be distinguished in event explanation: the question of the size of the probability attached to the event to be explained and the question of the change in the probability of the event brought about by a statistically relevant factor. I call these respectively the *size* debate and the *change* debate.

Three main positions have been taken in the size debate. The first is Hempel's: an explanation must show that the event to be explained has a high probability (at least greater than one-half); thus only high-probability events can be explained. I call this view *extreme elitism*. It has the consequence that heavy smoking alone cannot explain emphysema. Hempel would stand by this conclusion even if smoking were the only way to contract emphysema (compare Michael Scriven's

(1959) famous example of paresis), claiming that either some other predisposing cause of emphysema must be cited or, if there is none, we simply do not understand why the patient contracted emphysema.

The contrary view, that events with low probabilities can be explained, has two variants, one elitist and one egalitarian. According to the *moderate elitist*, the higher the probability of an event, the better it is explained by citing that probability. Low-probability events can be explained, then, but not as well as high-probability events. According to the *egalitarian*, events are equally well explained regardless of their probability.

Most philosophers are agreed that there are explanations of low-probability events to be found in science and so extreme elitism does not capture actual explanatory practice (though it might of course be regarded as a reformist proposal). It is more difficult to use scientific practice to adjudicate between moderate elitism and egalitarianism, since these views concur as to which events can be explained, disagreeing only as to how well they are explained. Michael Strevens (2000) argues that the history of the development of statistical physics favors the moderate elitist position.

There are also more philosophical, which is to say a priori, considerations that can be brought to bear on the debate. Some writers hold that to explain an event is to show why it occurred rather than not occurring; they see a tension between this doctrine and the egalitarian view that both the occurrence and the nonoccurrence of an indeterministically produced event can be explained equally well. (Salmon [1990, 178–179] formulates without endorsing the argument.)

Other writers hold that an explanation that enumerates accurately all the causes of the event it explains is perfect. A low-probability event is perfectly well explained, then, if the factors that determine its low probability are its only causes, which is presumably the case if (and only if) the low probability is of the irreducible variety, in particular if it is a quantum mechanical probability.

The change debate concerns the relationship between a factor's probabilistic impact on the event to be explained and its explanatory importance. Are factors that make a larger probabilistic impact explanatorily more important? Are factors that make a negative impact explanatorily important? As in the size debate, three positions can be distinguished. (Though I give the positions in the size and change debates similar names, you will see that they are distinct.)

On egalitarian views, the size of the impact does not affect a factor's explanatory value. Citation of a factor that makes a large difference to the probability of the event to be explained adds no more to an explanation than citation of a factor that makes only a small difference. To put it another way, appreciating the probabilistic role of a factor that makes a large difference no more illuminates the occurrence of the event to be explained than does appreciating the probabilistic role of a factor that makes virtually no difference. If heavy smoking quintuples your probability of contracting emphysema while having a wood-burning fire increases the probability by 1 percent, then an explanation of a person's emphysema that cites only his or her wood-burning fire is just as illuminating as an explanation that cites only his or her smoking. Despite the awkward sound of these consequences, many writers on probabilistic explanation, including the founder of the statistical relevance approach, Salmon, tend to egalitarianism.

A question that divides egalitarians concerns factors that lower the probability of the event to be explained. *Moderate egalitarians* hold that such factors are not explanatory, *extreme egalitarians* that they are. Salmon (1984, 46) has argued most explicitly for extreme egalitarianism, though even he forbids explanations that cite only probability-lowering factors. Humphreys is another staunch extreme egalitarian. Observe that any formal criterion for probabilistic explanation that requires the presentation of all factors that play a role in fixing the probability of the explanandum will render probability lowerers explanatorily relevant. Railton's and even Hempel's accounts of event explanation fit this description. On the other side of the divide lies *moderate elitism*, the view that the more a factor increases the probability of the event to be explained, the more it contributes to the event's explanation.

Although the resolution of the change debate would cast much light on the nature of probabilistic explanation—not least the problem of explanatorily irrelevant probability raisers, such as Petra's poisoning in the previous section—philosophers have not gone much further than laying out the issues. Such arguments as there are tend to reflect the arguments in the size debate. For example, on the causal approach to explanation you might think (as Humphreys argues) that a factor that decreases the probability of an event is a part of the causal history of that event and so qualifies alongside the event's probability raisers for a place in the explanation of the event. You will then tend to extreme egalitarianism in both the size and the change debates.

Or if you are a dispositionalist you might think that the probability of an event quantifies the "force" bringing the event about. The more force, the easier it is to understand the fact that the event occurred. High-probability events are therefore better explained, and greater contributions to the probability of an event are greater contributions to its explanation. Result: moderate elitism in the size and change debates. Although taking a position in the size debate does not force you to take any particular position in the change debate, then, there is an affinity to be found between same-named positions.

6. PROBABILITY AND DETERMINISM

Of the three varieties of probabilistic explanation described in section 1, only one paradigmatically involves the citation of an irreducible probability, such as a probability stipulated by the laws of quantum mechanics. The other two—explanation in which probability fills an epistemic gap, as in the case of emphysema, and explanation in complex systems where a probabilistic theory captures predisposing causes while abstracting from the details of particular initial conditions, as in statistical physics—work perfectly well even in deterministic systems.

A number of writers have cast doubt on the validity of probabilistic explanation in deterministic systems by way of the following two premises:

1. There are no physical probabilities in deterministic systems.
2. Where there are no physical probabilities, there can be no probabilistic explanation.

This is an incredible conclusion. After all, explanation in statistical physics and other high-level sciences that employ probability in a similar way is a well-established fact of scientific life, hardly ripe for overthrow by philosophers. The premises, then, merit a closer look.

The view that there can be no physical probabilities in a deterministic system has a certain intuitive appeal (Schaffer 2007). However, of the various philosophical accounts of the nature of physical probability, several allow the existence of such probabilities: the various versions of the frequency account, Karl Popper's propensity theory, and proposals to find the basis of certain systems' physical probabilities in the mathematical properties of the systems' dynamics (Hopf 1934; Strevens 2003, 2011; Abrams 2012).

Salmon, for example, adopts a version of the frequency theory to provide the metaphysical foundations for his statistical relevance account. But frequentism has fallen out of favor, and the view that probabilities can only be some kind of irreducible propensity has taken its place in the literature as the default view. This is the one interpretation of probability that disallows physical probability in deterministic systems. An apparent impasse, then: either the dominant metaphysics of probability must be discarded or perhaps augmented or the practice of probabilistic explanation eviscerated.

There is a third way, however. Premise 2 above, that there can be no probabilistic explanation without physical probability, is not as obviously true as it might seem.

One way around the premise is via accounts of probabilistic explanation that do not call directly on physical probability. Hempel's IS account is the preeminent example: what Hempel requires for explanation is in the first instance inductive, not physical, probability. (Hempel himself writes that the probabilistic laws in IS explanation must be based in real physical probabilities; however, he clearly has in mind a metaphysics of physical probability—presumably empiricist—on which there are probabilities wherever there are robust statistics.)

Not many contemporary philosophers would be willing to turn to the IS account to make room for probabilistic explanation in deterministic systems, however. There is a second option. It is generally accepted that, although the fundamental laws of nature are quite possibly indeterministic, deterministic explanation is legitimate when the fundamental probabilities are all close enough to zero and one. That is, deterministic explanation is possible in special circumstances in indeterministic systems. Perhaps, then, probabilistic explanation is possible in special circumstances in deterministic (or near deterministic) systems. More exactly, perhaps the apparatus of probabilistic explanation is especially well suited to capturing the explanatorily relevant aspects of certain systems that are at root deterministic and so, by some philosophers' lights, contain no physical probabilities.

Railton (1981) suggests that the probabilistic element of statistical physics functions in explanation to capture the robustness of the underlying physical processes. Since in section 1 I framed the presentation of probabilistic explanation in statistical physics in just these terms, Railton's view will sound familiar. The following discussion will flesh out the view, taking as a framework something somewhat stronger than Railton's suggestion. The framework is comprised of the following four posits (of which the first can be regarded, for the sake of the philosophical dispute, as uncontroversial):

1. In the domain of statistical physics (and, it might be added, many other areas), the kind of event you want to explain, such as the expansion of a gas to fill a vacuum, occurs given almost any set of initial conditions. Call this property *robustness*.
2. To understand a robustly produced event, you must grasp the reasons for the robustness of the underlying process.
3. To understand a robustly produced event, you need not appreciate the exact initial conditions that led to the event.
4. A probabilistic representation is especially good at capturing the explanatorily essential facts about robustness that are the subject of premise 2 while omitting the explanatorily uninteresting facts about the details of initial conditions that are the subject of premise 3.

Railton himself asserts premise 2, though without giving an account of the explanatory value of robustness; he would probably not endorse 3 and might or might not agree with 4. The following discussion explores more recent literature on robustness that better fits the framework.

The framework takes no position on the question of the existence of physical probabilities in systems to which its posits apply. You might believe that the robust elements of the systems' dynamics picked out by the probabilistic representation in some sense constitute genuine physical probabilities, you might deny it, or you might withhold judgment. The framework allows all three responses. A division of labor is therefore proposed: let philosophers of explanation articulate the explanatory value of probabilistic theories in systems with robust dynamics; let metaphysicians of probability decide whether the basis of such explanations includes genuine probabilities.

Event e is produced robustly. Why is understanding the basis of robustness explanatorily more important than understanding the exact initial conditions that led to e?

Frank Jackson and Philip Pettit (1992) argue for two modes of explanation, one in which the initial conditions are paramount and one in which the fact of robustness is what matters. In the latter mode the aim of explanation is to discover similarities between the causal process that actually produced e and the causal processes that produce e in nearby possible worlds, that is, roughly, the causal processes that might have produced e if the actual process had not. In a robust system these would-be producers of e are processes in the same system with different initial conditions. A good explanation, in Jackson and Pettit's second mode, points to what the processes have in common and overlooks their differences; it therefore cites the properties responsible for the system's robustness and ignores the particularity of initial conditions.

James Woodward (2003) sees an explanation of e as a body of information, in the form of a causal model, as to how to causally manipulate the occurrence of events like e. Information about robustness is useful for manipulation, but information as to exact initial conditions is not, since changing the initial conditions of a robust process is unlikely to change the outcome of the process.

Strevens (2004) proposes that an explanation of e specifies (some of) the factors that made a difference to the causal production of e. A causal detail fails to make a difference to e if a causal model for e that abstracts away from its presence—that does not specify how things are with respect to that detail—is sufficient to entail e, at least with a high probability. (Note that the entailment in question must mirror a real causal process.) When e is produced robustly, then, the details of the initial conditions that produced e are not difference makers, but the properties responsible for the robustness are. Unlike Jackson and Pettit and Woodward, then, Strevens holds that the explanatory relevance of robustness and the irrelevance of initial conditions are due to the role that these factors play in the actual causation of e.

There are a number of ways, as you can see, to account for the explanatory superiority of a probabilistic model that cites robustness and ignores initial conditions over a strictly deterministic model that cites the exact initial conditions. But what role does probability itself play in a robustness-based model?

One answer is that, in the case where only *almost all* initial conditions lead to the event e to be explained and where the conditions that do not lead to e are thoroughly mixed in with those that do, a deterministic model is for all practical purposes impossible. To specify the initial conditions sufficiently finely to rule out any possibility of conditions not producing e, you would have to disgorge such torrents of detail that you might as well simply state the actual initial conditions, defeating the purpose of a robustness-based model.

Another more interesting answer is that to understand the basis of robustness itself you must use the conceptual inventory of probability theory. This view is suggested by Strevens (2003, 2008). It is argued that the robustness-like properties of the processes described by statistical physics, evolutionary theory, and perhaps some of the social sciences are best understood by invoking all the apparatus of probability theory—probability densities, stochastic independence, the law of large numbers, the central limit theorem—regardless of whether there are "metaphysically real" physical probabilities at work in the system. On Strevens's proposal, an explanatory model for e that cites only the robustness of the process leading to e might be fully deterministic, yet it might also be probabilistic in the sense that it uses probabilistic concepts to derive and so to explain the robustness. The value of probabilistic explanation in deterministic systems is, then, not merely that probabilistic thinking is unavoidable but that it is invaluable, in some cases even where determinism reigns.

References

Abrams, M. 2012. "Mechanistic Probability." *Synthese* 187:343–375.
Achinstein, P. 1983. *The Nature of Explanation*. New York: Oxford University Press.
Coffa, A. 1974. "Hempel's Ambiguity." *Synthese* 28:141–163.
Fetzer, J. 1974. "A Single Case Propensity Theory of Explanation." *Synthese* 28:171–198.
Gluck, S., and S. Gimbel. 1997. "An Intervening Cause Counterexample to Railton's DNP Model of Explanation." *Philosophy of Science* 64:692–697.
Hempel, C. 1965. "Aspects of Scientific Explanation." In *Aspects of Scientific Explanation and Other Essays in the Philosophy of Science*, 331–496. New York: Free Press.
Hempel, C., and P. Oppenheim. 1948. "Studies in the Logic of Explanation." *Philosophy of Science* 15:135–175.
Hopf, E. 1934. "On Causality, Statistics, and Probability." *Journal of Mathematics and Physics* 13:51–102.
Humphreys, P. 1989. *The Chances of Explanation*. Princeton, NJ: Princeton University Press.
Jackson, F., and P. Pettit. 1992. "In Defense of Explanatory Ecumenism." *Economics and Philosophy* 8:1–21.
Jeffrey, R. 1971. "Statistical Explanation vs. Statistical Inference." In *Statistical Explanation and Statistical Relevance*, by W. Salmon, R. Jeffrey, and J. Greeno, 19–28. Pittsburgh: University of Pittsburgh Press.
Menzies, P. 1996. "Probabilistic Causation and the Pre-emption Problem." *Mind*, n.s., 105:85–117.
Railton, P. 1978. "A Deductive-Nomological Model of Probabilistic Explanation." *Philosophy of Science* 45:206–226.
———. 1981. "Probability, Explanation, and Information." *Synthese* 48:233–256.
Salmon, W. 1971. "Statistical Explanation." In *Statistical Explanation and Statistical Relevance*, by W. Salmon, R. Jeffrey, and J. Greeno, 29–88. Pittsburgh: University of Pittsburgh Press.
———. 1984. *Scientific Explanation and the Causal Structure of the World*. Princeton, NJ: Princeton University Press.
———. 1990. *Four Decades of Scientific Explanation*. Minneapolis: University of Minnesota Press.

Schaffer, J. 2007. "Deterministic Chance?" *British Journal for the Philosophy of Science* 58:113–140.

Scriven, M. 1959. "Explanation and Prediction in Evolutionary Theory." *Science* 30:477–482.

Strevens, M. 1998. "Inferring Probabilities from Symmetries." *Noûs* 32:231–246.

———. 2000. "Do Large Probabilities Explain Better?" *Philosophy of Science* 67:366–390.

———. 2003. *Bigger Than Chaos: Understanding Complexity through Probability*. Cambridge, MA: Harvard University Press.

———. 2004. "The Causal and Unification Approaches to Explanation Unified—Causally." *Noûs* 38:154–176.

———. 2008. *Depth: An Account of Scientific Explanation*. Cambridge, MA: Harvard University Press.

———. 2011. "Probability out of Determinism." In *Probabilities in Physics*, ed. C. Beisbart and S. Hartmann, 339–364. New York: Oxford University Press.

Woodward, J. 2003. *Making Things Happen: A Theory of Causal Explanation*. New York: Oxford University Press.

Chapter 3

Laws of Nature

MARC LANGE

1. INTRODUCTION

The laws of nature have traditionally been contrasted with facts of two other kinds: facts that absolutely could not have been otherwise (such as logical, conceptual, mathematical, and metaphysical truths), on the one hand, and accidental facts, on the other hand. Typical examples of accidents in the philosophical literature are that all of the coins in my pocket today are silver (Goodman 1983, 10) and that all solid gold cubes are smaller than a cubic mile (Reichenbach 1947, 368; 1954, 10). Like the accidents and unlike the logical, conceptual, mathematical, and metaphysical truths, the laws are traditionally considered to be contingent truths that science discovers empirically. However, an accident just happens to obtain. A gold cube larger than a cubic mile could have formed, but as it happened (let us suppose), enough gold to form one never came together. In contrast, it is no accident that an electrically insulating copper object (for example) never formed. The natural laws prohibit such a thing. Events *must* conform to the laws of nature, whereas accidents are mere coincidences. Like the logical, conceptual, mathematical, and metaphysical truths, the laws possess a kind of necessity—though, being contingent, the laws cannot be *as* necessary as the facts that absolutely could not have been otherwise. The laws' weaker sort of necessity is usually called "nomic" or "physical" necessity.

The natural laws' paradoxical-sounding status as "contingent necessities"—as lying somehow between the absolute necessities and the accidents—raises a host of difficult questions. Some philosophers study the distinctive roles that laws play in scientific reasoning.[1] Philosophers also try to figure out what the laws of nature *are*: what it is about the laws that enables them to play these special roles, and what it is about the universe in virtue of which it is governed by certain laws. These questions bear upon more general metaphysical issues.

Other philosophers argue that the traditional picture sketched above is mistaken in various respects. They may argue, for example, that science fails to draw a sharp distinction between laws and accidents. Alternatively, they may argue that laws are absolutely necessary (though known a posteriori) rather than contingent truths. Each of these views denies that the laws lie between the absolute necessities and the accidents. Some philosophers investigate whether there are laws of "inexact" (or "special") sciences or whether the only genuine laws are those of fundamental physics. If there are laws of inexact sciences, then these sciences are autonomous: they have their own natural kinds and they supply scientific explanations that are irreducible to those given by fundamental physics.[2]

2. LAWS AND COUNTERFACTUALS

Although all gold cubes are in fact (let us suppose) smaller than a cubic mile, I dare say that there would have been a gold cube exceeding one cubic mile had Bill Gates wanted such a cube constructed. But even if he

---

1. Although some laws (such as Heisenberg's uncertainty "principle," Archimedes's "principle," the "axioms" of quantum mechanics, and Maxwell's "equations") are usually not called "laws," they play the roles characteristic of laws in scientific reasoning. Laws are distinguished by what they *do*, not by what they are called. Likewise, some facts commonly called "laws" are really accidental. For example, Kepler's first "law" of planetary motion (that all of the planets trace approximately elliptical orbits with the sun at a focus) presupposes that the planets' masses happen to be negligible compared to the sun's.
2. Fodor 1974 is an influential argument for autonomy.

had wanted to have an electrically insulating copper object constructed, all copper objects would still have been electrically conductive, since events are obliged to conform to the natural laws. In other words, the laws govern not only what actually happens, but also what would have happened under various circumstances that did not actually happen. Accordingly, the laws are often said to "support" counterfactual conditionals (or "counterfactuals", for short), that is, to "underwrite" facts expressed by statements of the form "Had p been the case, then q would have been the case." (This is commonly symbolized as $p \rightarrow q$.)

Talk of laws "supporting" counterfactuals can be interpreted in various ways. It can be given an epistemological gloss: to figure out which counterfactuals are true (e.g., which terrestrial conditions would have been different had Earth's rate of rotation on its axis been half as great), we consult the laws of nature. Alternatively, the slogan that laws "support" counterfactuals could be given an ontological gloss: the law that gravitational forces diminish with the square of the distance is part of what makes it true that had two point masses been twice as far apart, their mutual gravitational attraction would have been one-fourth as great. To avoid presupposing that the laws are ontologically or epistemically prior to the counterfactual truths, we could say simply that the laws would have been no different (whereas some accidents would have been different) had p been the case, for any counterfactual supposition p that is "nomically possible," that is, logically consistent with all of the laws. (Among those who have defended ideas along these lines are Jonathan Bennett [1984], John W. Carroll [1994], Roderick M. Chisholm [1946, 1955], Nelson Goodman [1947, 1983], Paul Horwich [1987], Frank Jackson [1977], J. L. Mackie [1962], John L. Pollock [1976], and P. F. Strawson [1952].) Obviously, if p is logically *in*consistent with the laws, then had p been the case, the laws would have been different; for example, had the electromagnetic force been somewhat stronger, then the carbon nucleus would not have held together against the mutual electrostatic repulsion of its protons. (Such a counterfactual is called a "counterlegal.")

However, there are various reasons to worry about whether this idea adequately captures the difference between laws and accidents in their relation to counterfactuals. To begin with, does "the laws would have been no different" mean that the laws would still have been *true* or that the laws would still have been *laws*? To use the laws to infer what would have happened under various unrealized circumstances presupposes only that the laws would still have been true. But unless the laws would still have been laws, not accidental truths, they cannot tell us about the *counterfactual conditionals* that would have held under various unrealized circumstances. (The nested counterfactual $p \rightarrow (q \rightarrow r)$ is not in general equivalent to $(p\&q) \rightarrow r$. For example, "Had the object been pure rubber, then had it been pure copper, it would have been electrically conductive" is not equivalent to "Had the object been pure rubber and pure copper, it would have been electrically conductive.") For instance, we believe that had Jones missed his bus to work this morning, the laws of nature would still have been laws, and so had Jones then clicked his heels and made a wish to arrive instantly at his office, his wish would not have come true.

By the same token, for a counterfactual supposition p to be "nomically possible" (and so to fall within the scope of the above idea), does it suffice for p to be logically consistent with the conjunction of every m where it is a law that m, or must p be logically consistent with the conjunction of all truths of the form "It is a law that m" together with all truths of the form "It is not a law that m"? For example, suppose p is that *it is not a law* that momentum is conserved. Presumably, p is logically consistent with momentum being conserved in each interaction that occurs (together with every other m where it is a law that m). But presumably, momentum would not always have been conserved, had momentum conservation not been required by law.

Let us try to refine this means of distinguishing laws from accidents by thinking of the laws as belonging to a hierarchy, where the facts on one rung of the hierarchy are facts about what is governing the facts on the rung below. High in the hierarchy are facts such as that it is a law that all laws are Lorentz invariant and other such symmetry principles. The facts on that rung ("metalaws") are facts about what it is that other laws are governed by.[3] One rung below are the facts thereby governed, such

---

3. Whereas Eugene Paul Wigner (1967, 43) places symmetries in the nomic hierarchy, Bas van Fraassen (1989) deprecates laws in favor of symmetries as capturing the sense in which different cases are essentially alike.

as the fact that it is a law that all electrons have negative electric charge and the fact that it is not a law that all gold cubes are smaller than a cubic mile. These facts are concerned not with what governs other *laws*, but rather with what governs the facts on the rung below, which is the lowest rung of the hierarchy. The bottom-rung facts are not about what governs some other facts. Let us call these bottom-rung facts "nonnomic" and reserve lowercase letters for claims that, if true, state nonnomic facts. Whereas p (e.g., "My shirt is blue," "All copper objects are electrically conductive") is a nonnomic claim, "It is [not] a law that p" is not; it is one rung higher. By confining our attention to counterfactual suppositions of the form "Had p" (e.g., "Had there existed a gold cube larger than a cubic mile," "Had there existed an electrically insulating copper object"—but not "Had it not been *a law* that momentum is conserved"), we can try to distinguish the nonnomic facts that are laws (i.e., the n's where it is a law that n) from those that are accidents, as follows:

n follows from the laws if and only if

- n would still have been true, had p been the case (i.e., $p \to n$), for every p that is logically consistent with the conjunction of every m where it is a law that m,[4] and
- likewise all nested counterfactuals $p \to (q \to n)$, $r \to (p \to (q \to n))$, and so forth are true where p is logically consistent with the conjunction of every such m, and so is q, and so is r, and so forth.

Let us call this principle "nomic preservation." Does it capture the laws' distinctive relation to counterfactuals?

One reason to worry about nomic preservation is that counterfactuals are notoriously context sensitive. In certain contexts the counterfactual "Had Caesar been in command in the Korean War, he would have used the atomic bomb" is true (e.g., in contexts where we are concerned with Caesar's military style), whereas in other contexts (where

---

4. By "logically consistent," an advocate of this principle would have to mean "in a broad sense": p must be logically, mathematically, conceptually, and metaphysically compossible with every m where it is a law that m.

we are concerned instead with Caesar's military knowledge) "he would have used catapults" is true (Quine 1960, 222). Since what is preserved under a counterfactual supposition and what is allowed to vary depends greatly upon our interests in entertaining the supposition, perhaps there exist contexts in which the laws would not still have held under some nomic possibility p. On the other hand, perhaps it is constitutive of lawhood (and hence true irrespective of context) that the laws would still have been true under any nomic possibility p.

Here is another reason to worry about nomic preservation (Lewis 1986, 171). When we consider what would have happened had Jones missed his bus to work, we typically consider how the future would have unfolded under this altered circumstance with the past just as it actually was. (For example, in view of Jones's actual exemplary past record of arriving to work on time, his boss might easily have forgiven his lateness on this occasion.) But if the laws are deterministic and Jones did not actually miss his bus, then the laws plus the actual past is logically inconsistent with Jones missing his bus. Apparently, to preserve the actual past and yet accommodate Jones missing his bus, the "closest possible world" where Jones misses his bus must contain a violation of the actual laws of nature—what David Lewis (1986, 48) calls "a small miracle." So the laws of nature would have been different, had Jones missed his bus. But this result strikes many as counterintuitive and contrary to scientific practice. Perhaps the counterfactual supposition "Had Jones missed his bus" evokes not an entire possible world, but only a partial world history (a "nonmaximal situation") that includes certain events in the past (such as Jones's exemplary on time record) but does not include the "miracle," since our concern in this context is with the consequences of Jones missing his bus, not with how Jones managed to miss his bus. No event in this partial history violates the actual laws, so they are preserved. (If we switch to a context where we *are* concerned with how Jones came to miss his bus, then the complete actual past is not preserved.)

Even setting these challenges aside, we ought to find it worrisome that nomic preservation uses consistency with the laws to delimit the range of counterfactual suppositions under which a fact must be preserved for it to qualify as following from the laws. It is trivial that no accident is preserved under all counterfactual suppositions that are consistent with the laws (since no accident a would still have held under the supposition $\sim a$); the range of counterfactual suppositions under consideration has plainly been designed expressly to suit the laws. Although nomic preservation may set the laws apart from the accidents, it leaves unexplained why preservation under this particular range of counterfactual suppositions is especially important. For nomic preservation to tell us why the laws are special, we would first need to know what is special about being preserved under the counterfactual suppositions that are logically consistent with the laws, and so we would already need to know why the laws are special. The root of the problem is that the laws are mentioned on *both* sides of the "if" in the definition of nomic preservation.

3. STABILITY AND NOMIC NECESSITY

In view of such difficulties, some philosophers (see, e.g., Mitchell 2000; Woodward 2001) have contended that science draws no sharp distinction between laws and accidents. Rather, different facts simply have different ranges of invariance under counterfactual perturbations. Indeed for nearly any accident, there are *some* counterfactual suppositions under which it is preserved; for example, all gold cubes would still have been smaller than a cubic mile even if I had worn a different shirt today. Some accidents have considerable resilience under counterfactual twiddles. For instance, my car's maximum speed on a dry, flat road stands in a certain relation to the distance of my car's gas pedal from the floor, and though this relation is not traditionally regarded as a law (since it reflects accidental features of my car's engine), it has "invariance with respect to certain hypothetical changes" (Haavelmo 1944, 29), such as "Had I depressed the gas pedal further toward the floor."

Perhaps, however, there remains a way to draw a sharp, significant distinction between laws and accidents. In nomic preservation, the set of laws picks out the broadest range of counterfactual suppositions convenient to itself: the suppositions that produce no logical contradiction when combined with every m where it is a law that m. What if we extend the same courtesy to a set containing accidents, allowing it to pick out the broadest range of counterfactual suppositions convenient to itself: the suppositions that produce no logical contradiction when combined with every member of the set? Take, for example, a logically closed set of truths that includes the fact that all gold cubes are smaller than a cubic mile. The set's members would not all still have been true had Bill Gates wanted to have a large gold cube constructed. So for the set's members all to be preserved under every counterfactual supposition that is logically consistent with all of them, the set must contain the fact that Gates never wants to have a large gold cube constructed; the counterfactual supposition that he wants to is then logically inconsistent with a member of the set. However, presumably, had Melinda Gates (Bill's wife) wanted to have a large gold cube constructed, then Bill would also have wanted one built. So having included the fact that Bill Gates never wants to have a large gold cube constructed, the set must also include the fact that Melinda Gates never does either in order for all of the set's members still to have been true under any counterfactual supposition with which every member is logically consistent.

Such a set must be very inclusive. Suppose, for example, that the set omits the accident that all of the apples now on my tree are ripe. Then here is a counterfactual supposition that is logically consistent with every member of the set: had either some gold cube exceeded one cubic mile or some apple on my tree not been ripe. Under this supposition, there is no reason why the gold-cube generalization (which is a member of the set) takes precedence in every conversational context over the apple generalization (which we supposed not to be a member of the set). So it is not the case that the gold-cube generalization is preserved (in every context) under this counterfactual supposition. Hence, the set must include the apple generalization, too, if the set is to be invariant under every counterfactual supposition that is logically consistent with its members. The upshot is that if a logically closed set of truths includes an accident, then it must include every accident if it is to be invariant under every counterfactual supposition that is logically consistent with every member of the set.

But according to nomic preservation, the laws' logical closure exhibits exactly this kind of invariance. So we can now specify the laws' distinctive behavior under counterfactual perturbations without using the laws to specify the relevant range of counterfactual suppositions: the laws are special in forming a set that is invariant under the broadest range of counterfactual suppositions that the set itself picks out. In other words, take a logically closed set Γ of truths m that is not the empty set. Define:

Γ is *stable* if and only if for each member n of Γ

- n would still have been true, had p been the case (i.e., $(p \to n)$), for every p that is logically consistent with every member m of Γ, and
- likewise all nested counterfactuals $p \to (q \to n)$, $r \to (p \to (q \to n))$, and so forth are true for every q that is logically consistent with every member of Γ, every p that is logically consistent with every member of Γ, every r that is logically consistent with every member of Γ, and so forth.

The set containing exactly the laws n (together with the logical, mathematical, conceptual, and metaphysical necessities and all of the logical consequences thereof) is stable, whereas no set Γ containing some accidents is stable except the set of all truths n, which is trivially stable (since no counterfactual supposition q is logically consistent with all of its members).

A set is stable if and only if its members are jointly invariant under as broad a range of counterfactual suppositions as they could jointly be (since they could not all be preserved under a supposition with which one of them is logically inconsistent). So the laws collectively are maximally invariant. In virtue of this behavior, the laws possess a kind of necessity. Intuitively, necessity involves being the case no matter what, in the broadest possible sense of "no matter what." No necessity of any kind is possessed by an accident, even an accident (such as my car's pedal-speed function) that would still have held under many counterfactual suppositions, because a stable set cannot be constructed around an accident (except trivially). In contrast, the logical, conceptual, mathematical, and metaphysical truths form a stable set and so possess a kind of necessity. Since this set is a proper subset of the stable set that also includes all of the laws, the range of counterfactual suppositions under which it is preserved, in being stable, contains and extends beyond the range of counterfactual suppositions under which the laws are preserved in being stable. Thus, the notion of stability allows us to capture the sense in which the laws are "contingent necessities": necessary but not *as* necessary as the logical, conceptual, mathematical, and metaphysical necessities and lying between them and the accidents.

Furthermore, if it is a law that n exactly when n belongs to a nontrivially stable set, then the laws are not only still *true*, but also still *laws* under every counterfactual supposition p that produces no contradiction when combined with the laws m. The nested counterfactuals in the definition of "stability" ensure that under any such supposition, the laws would still have formed a nontrivially stable set. That is, they ensure that the laws' distinctive invariance under counterfactual suppositions is itself "no accident."

The laws' stability also reveals a sense in which the laws form a system—an integrated whole. Each member of the set spanned by the laws helps delimit the range of counterfactual suppositions under which every other member must be preserved for the set to be stable. Lawhood is thus a collective affair rather than an individual achievement.

For any two stable sets, one must be a proper subset of the other (Lange 2005). Perhaps there are several nontrivially stable sets in fundamental physics: one set spanned by the fundamental dynamical laws and a more inclusive set spanned by these dynamical laws plus the laws characterizing nature's basic constituents (e.g., the particular kinds of "particles" and "forces" there are). Furthermore, perhaps the existence of nontrivially stable sets other than the set spanned by the familiar laws accounts for the ambiguous modal status of certain facts (cf. Sklar 1991). For example, although the second law of thermodynamics seems to function as a law in scientific reasoning, it requires that the initial microconditions of closed systems in a given initial macrocondition tend to be distributed in a certain manner among the microconditions compatible with that macrocondition—and this distribution appears accidental. (For example, for the perfume from recently opened bottles to be more likely to spread quickly throughout the room than to remain in the bottles, the perfume and air molecules must tend not to be "coordinated" so that whenever a perfume molecule threatens to escape from a bottle, another molecule happens to come along and knock it back inside.)

Perhaps one nontrivially stable set is spanned by the fundamental laws of physics, whereas another also includes the second law of thermodynamics along with the probability distribution governing various initial microconditions. The natural laws would then form several strata between the absolute necessities and the accidents.

4. LAWS AND SCIENTIFIC EXPLANATIONS, INDUCTIVE CONFIRMATION, AND NATURAL KINDS

Besides standing in different relations to counterfactuals, laws and accidents have traditionally been thought to display other differences as well. Because of their necessity, laws have been ascribed an explanatory power that accidents lack (Hempel 1965, 339). For example, a certain powder burns with yellow flames, rather than flames of another color, because the powder is a sodium salt and it is a law that all sodium salts, when ignited, burn with yellow flames. (This law, in turn, is explained by more fundamental laws.) The powder *had to* burn with a yellow flame considering that it was a sodium salt—and that "had-to-ness" expresses the law's distinctive variety of necessity; being a sodium salt, the powder would have burned with a yellow flame no matter what. In contrast, we cannot explain why my wife and I have two children by citing the regularity that all of the families living on our block have two children, since this regularity is accidental. Had a childless family tried to move onto our block, it would not have encountered an irresistible opposing force.

Since we believe that it would be merely coincidental if all of the coins in my pocket today turned out to be silver, our discovery that the first coin I withdraw from my pocket today is silver fails to justify raising our confidence in the hypothesis that the next coin to be examined from my pocket today will also turn out to be silver. Likewise, since we believe that it would be an accident if all US presidents elected in years ending in "0" died in office, we regard our discovery that Warren G. Harding (elected in 1920) died in office as failing to confirm that William H. Harrison (elected in 1840) died in office. A candidate law is confirmed differently (according to Armstrong 1983, 4; Black 1967,

171; Braithwaite 1927, 467; Dretske 1977, 256–260; Goodman 1983, 20, 69–73; Kim 1992, 11–13; Kneale 1952, 45, 52; Mill 1884, 230–231; Moore 1962, 12; Peirce 1934, 55; Reichenbach 1947, 359–361; Scheffler 1981, 225–227; Strawson 1952, 199–200). That one sample of a given chemical substance melts at 383K (under standard conditions) confirms, for every unexamined sample of that substance, that its melting point is 383K (in standard conditions).

This difference in inductive role between candidate laws and candidate accidents seems related to the fact that laws, unlike accidents, express similarities among things that reflect their belonging to the same "natural kind" (Hacking 1991). The electron, the emerald, and the electromagnetic force are all natural kinds, and accordingly there are laws of the form "All electrons…," "All emeralds…," and "All electromagnetic interactions.…" In contrast, the families on my block do not form a natural kind, nor do the gold cubes. Accordingly, there are no laws covering exactly these classes.

If the laws cover the natural kinds, then some contingent logical consequences of laws are not laws (although they possess nomic necessity), since they do not cover natural kinds (Fodor 1974). Consider, for example, "All things that are emeralds or electrons are green or negatively charged" and "All gold cubes are electrically conductive." The former tries to unify too much and the latter fails to unify enough. This distinction between laws and nomically necessary nonlaws seems to figure in scientific practice, as when nineteenth-century chemists took it to be a law that all noncyclic alkane hydrocarbons differ in atomic weight by multiples of 14 units and a law that all nitrogen atoms weigh 14 units but a mere coincidence that all noncyclic alkane hydrocarbons differ in atomic weight by multiples of the weight of the nitrogen atom. (See also Mill 1884, 229–230.) However, some philosophers regard all contingent logical consequences of laws as laws (Braithwaite 1953, 305; Davidson 1990, 218; Hempel 1965, 346–347; Lewis 1983, 367–368; Nagel 1961, 60).

Although candidate laws seem to differ from candidate accidents in the way that they are confirmed, a candidate accident can sometimes have its predictive accuracy confirmed, as when we confirm that the other apples now on my tree are ripe by tasting one of the apples from my tree and finding it to be ripe or when we regard an instance as confirming the predictive accuracy of a hypothesis regarding my car's pedal-speed relation. Some philosophers (see, e.g., Loewer 1996; Sober 1988) have accordingly denied that hypotheses believed to be "law-like" (i.e., to be laws if true) are confirmed differently from hypotheses believed to be accidental if true. Some philosophers (see, e.g., van Fraassen 1989, 134–138, 163–169) have concluded that the concept of a natural law should not figure in our rational reconstruction of scientific reasoning.

On the other hand, the sort of confirmation to which hypotheses believed law-like alone are susceptible may need to be characterized more precisely than merely as confirmation of the predictive accuracy of a hypothesis. Predictions regarding actual unexamined cases and claims regarding counterfactual (and perforce unexamined) cases are confirmed together. Suppose we confirm h "inductively" only when we confirm not merely that h is actually true but also that h would still have been true under various unrealized circumstances: those logically consistent with all of the hypotheses that we are confirming inductively. Then each hypothesis that is undergoing inductive confirmation limits the range of unexamined cases over which the others are being inductively projected. All that we thereby confirm will turn out to be true if and only if the nomic necessities (of a certain stratum) are exactly the contingent logical consequences of the hypotheses h that we are confirming inductively. (That follows from the connection between lawhood and stability.) Accordingly, the laws are plausibly the best hypotheses for us to confirm inductively, and so to confirm h inductively, we must believe that h may be nomically necessary.

Some accidents (such as my car's pedal-speed relation) seem to have roughly the same explanatory significance as laws. Furthermore, although Carl G. Hempel presumes that facts can be explained only by other facts, population biology commonly uses idealized models (e.g., models supposing some population to be infinitely large) to explain changes through the generations in the relative frequencies of various genotypes. Economics likewise uses models that are not strictly true. The ideal gas law (which supposes molecules to be point particles), Snell's law (which supposes perfectly optically isotropic media), and Hooke's law (which presupposes that there are no nonlinear forces) are all used to explain the behavior of actual systems that are not ideal. Accordingly, some philosophers (see Cartwright 1983; Giere 1988) have suggested that laws are false as claims about the world but accurately describe certain scientific models that explain the behavior of certain real systems. (I shall return to such views.)

5. WHAT *IS* A LAW OF NATURE?

To explain the special roles that laws of nature play in scientific practice, philosophers have proposed various accounts of what laws are supposed to be. The two most popular contenders are (i) Lewis's (1973, 72–77; 1983; 1986; 1999) "best system" account, of which Lewis finds hints in John Stuart Mill and Frank P. Ramsey (so the proposal is sometimes called the Mill-Ramsey-Lewis or MRL account), and (ii) the "nomic necessitarian" accounts proposed by Fred I. Dretske (1977), D. M. Armstrong (1978, 1983, 1997), and Michael Tooley (1977, 1987), often called the DTA account.

According to Lewis, the facts about which truths are laws and which are not "supervene" on the facts regarding spacetime's geometry and the spatiotemporal mosaic of instantiations of certain elite properties. That is, two possible worlds[5] cannot differ in their laws without differing in their mosaics of elite-property instantiations. These elite properties are the properties meeting the following conditions:

- They are perfectly natural—unlike, for instance, the "disjunctive" property of being an emerald or greater than three inches long. That is, they are among the "sparse" properties—nongerrymandered ones, not mere shadows of predicates.
- They are categorical—that is, "Humean": they involve no modalities, propensities, chances, laws, counterfactuals, or dispositions.

---

5. Lewis restricts the scope to possible worlds containing no instances of "alien properties" (**Lewis 1986, p. 10**). Unlike many philosophers who believe that the laws supervene, Lewis does not regard the laws' supervenience as a metaphysical necessity.

- They are qualitative (i.e., nonhaecceitistic) in the sense that they do not involve the property (a "haecceity"), which a given thing intrinsically possesses (according to some philosophers), of being the particular individual thing that it is.
- They are possessed intrinsically by spacetime points or occupants thereof.[6]

Also supervening on the Humean mosaic are the facts about single-case objective chances, such as this atom's having a 50 percent chance of undergoing radioactive decay in the next 102 seconds.

Consider the various deductively closed systems containing only (though perhaps not all of the) truths regarding instantiations of elite properties and claims regarding the objective chances at various times that certain elite properties will be instantiated at later times.[7] These systems compete along three dimensions:

1. informativeness (in excluding or in assigning chances to various arrangements of elite-property instantiations),
2. simplicity (e.g., in the number of axioms and the order of polynomials therein, as expressed in terms of natural properties, spacetime relations, and chances), and
3. fit (which is greater insofar as the actual course of elite-property instantiations receives higher probability).

These three criteria stand in some tension. Greater informativeness can be achieved by adding facts to the system, which often (though not always) brings a loss of simplicity. Likewise, if property P is instantiated at time t_2, then by adding to a system the claim that c is the chance at t_1 of

---

6. This requirement could perhaps be relaxed in view of the holism suggested by quantum mechanics, where the state of an "entangled" pair of quantum particles fails to supervene on separate states of the two particles. See Loewer 1996.
7. Lewis also requires that any system eligible for the upcoming competition not say p without also saying that p never had any chance of not coming about. In this (ad hoc?) manner, Lewis arranges for "coherence" between the chances and the deterministic laws. That is, without this condition, Lewis's account would permit "All F's are later G'" to be a law though a given F at some moment has a non-zero chance of later being non-G.

P being instantiated at t_2, we may add informativeness—though not as much as we would, had we added that P is instantiated at t_2. We may also add fit—though not as much as we would, had we assigned P's instantiation a greater chance.

Perhaps some single system is by far the best on balance in meeting these three criteria. Perhaps which system "wins" the competition is relatively insensitive to any arbitrary features of our sense of simplicity or our rate of exchange among the three criteria.[8] In that case, the laws of nature are the contingent generalizations belonging to the best system, and the facts about the chances at a given moment are whatever the best system (and the history of elite-property instantiations until that moment) entails them to be. Roughly speaking, the addition of an accident to the best system would not contribute enough strength to outweigh the resulting loss of simplicity.

Some philosophers regard Lewis's account as having the virtue of using only Humean resources to distinguish between laws and accidents.[9] Unlike Humean accounts according to which the laws are distinguished from the accidents by the attitudes we bear toward them or the use we make of them (e.g., Goodman 1983, 20–22), Lewis construes the distinction between laws and accidents as objective.[10] Lewis's account also nicely accommodates "uninstantiated" (i.e., "vacuous") laws. Suppose we replace Coulomb's law in the best system by a generalization that agrees with Coulomb's law except in the case of a point body of exactly 1.234 statcoulombs at exactly 5 centimeters from a point body of exactly 6.789 statcoulombs. If no such pair of bodies ever exists, then the replacement generalization is true, just like Coulomb's law. However, it

---

8. If nature is "unkind" to us in that which system qualifies as "best" is not robust in this way, then "there would be no very good deservers of the name of laws. But what of it? We haven't the slightest reason to think the case really arises" (Lewis 1999, 233).
9. However, Lewis's account presupposes that certain properties are "natural"; see especially Lewis 1983.
10. Humean (or "regularity") accounts are sometimes divided into "naive" and "sophisticated" varieties. The former identify the laws as all and only the contingent regularities (or perhaps just those regularities that can be expressed without referring to any particular time, place, object, etc.). A "naive" regularity account fails properly to distinguish laws from accidents. Lewis's and Goodman's are both "sophisticated" regularity accounts.

is not as simple as Coulomb's law, since it treats this hypothetical pair of bodies as a separate case. So the best system contains Coulomb's law, replete with uninstantiated cases.

That the laws supervene on the Humean mosaic appears to render the laws epistemically more accessible and metaphysically less mysterious than they would otherwise be. On the other hand, insofar as the laws are like the rules of the game that the entities inhabiting the universe play, it seems doubtful that the laws supervene on the Humean mosaic, just as the rules of chess fail to supervene on the actual moves that are made in a given game.[11] (If no one castles in the course of the game, for instance, then the actual moves made in the game fail to reveal whether castling was permitted.) There seem to be two possible worlds with identical Humean mosaics, but whereas in one world it is a law that all F's are G, this regularity is accidental in the other world; there it is a law that all F's have a 99.99 percent chance of being G. Or suppose there had been nothing in the entire history of the universe except a single electron moving uniformly forever.[12] Lewis's account dictates that the laws would then have included "At all times, there exists only one body." But according to nomic preservation, the laws of nature would have been no different had there been nothing but a single electron. The universe's initial conditions alone would have been different. However, Lewis might reply, had there been nothing but a lone electron, then a great many actual laws would have been uninstantiated. They would then have been true trivially, but what would have made them laws? Furthermore, as I mentioned earlier, Lewis rejects nomic preservation on the grounds that it demands (in a deterministic universe) that had Jones missed his bus to work this morning, then the universe's state at every past moment would have been different. On the other hand, if Coulomb's law would not still have been a law had there been nothing but a lone

---

11. "The chess-board is the world, the pieces are the phenomena of the universe, the rules of the game are what we call the laws of Nature" (Huxley 1893, 3:82).
12. For the sake of the example, suppose that the electron is not accompanied by an electromagnetic field; rather electromagnetic interactions are action at a distance. Considerably more sophisticated arguments against the laws' supervenience on the Humean mosaic can be found in Carroll 1994, 60–68 and Tooley 1977, 669–672. For replies, see Beebee 2000; Loewer 1996; and Roberts 1998.

electron, then why is it the case that had there been nothing but a lone electron, then had there been another electron, the force between the two electrons would have accorded with Coulomb's law? Lewis would presumably reject this nested counterfactual as relying upon misleading, non-Humean intuitions.

Lewis takes the laws as arising "from below," out of the Humean mosaic. In contrast, DTA emphasizes that the laws "govern" the universe and so takes the laws as imposed on the Humean mosaic "from above"; the laws are facts over and above the facts they govern. According to DTA, the laws are irreducible, contingent relations of "nomic necessitation" among universals. That is to say, laws are not regularities among property instantiations but instead are relations among properties, such as the property of being an electron and the property of being negatively charged.[13] Such a relation's holding metaphysically imposes a corresponding uniformity among Humean facts. Whereas DTA holds that the actual relations of nomic necessitation among universals would still have held had Jones missed his bus to work this morning, Lewis believes that (in a deterministic world) this counterfactual supposition requires a "small miracle" (a single localized violation of the actual laws) for it to be accommodated in the least disruptive fashion, and so the laws would have been different had Jones missed his bus. On Lewis's view, the fact that the laws are not held "sacred" under such counterfactual suppositions is best explained by an account of law according to which there is no great metaphysical gulf separating laws from accidents. On DTA's view, the laws' distinctive powers to explain, to be confirmed inductively, and to support counterfactuals cannot be accounted for if laws are merely regularities, even regularities belonging to Lewis's best system. They must be facts of an ontologically more exalted kind: relations among universals.

Lewis's laws are ontologically posterior to the property instantiations in the Humean mosaic and so apparently cannot genuinely explain those events. Lewis (1999, 232) replies that a scientific "explanation"

---

13. Whereas Tooley countenances uninstantiated universals, Armstrong's "naturalism" demands that all universals be instantiated and so requires a more complicated account of uninstantiated laws.

explains precisely by unifying and systematizing—that is, by placing the explanandum in relation to the best system. Moreover, Lewis contends that this construal of scientific explanation is no more question begging than DTA's view that laws explain by virtue of special, explanatorily potent relations among universals. Just because this relation is called "nomic *necessitation*" does not mean that if F-hood and G-hood are related by such a relation, then an instantiation of F-hood really *compels* an instantiation of G-hood—unless DTA makes the ad hoc stipulation that "compels" here just means that the universals stand in a relation of "nomic necessitation" (Lewis 1983, 366; 1986, xii; van Fraassen 1989, 98).

Some philosophers believe that even contingent relations among universals are not ontologically exalted enough to account for the roles that laws characteristically play. On the "causal powers" or "essentialist" account, a rival to both MRL and DTA that is gaining increased attention at the time of this writing (Bigelow et al. 1992; Ellis 1999, 2001; Lierse 1999; Shoemaker 1980, 1998; Swoyer 1982; and, somewhat earlier, Harré and Madden 1975), the laws are not contingent. Rather, they are metaphysical necessities (that we ascertain a posteriori). According to essentialism, it is part of electric charge's essence (or nature) that it involves the causal power (or capacity, propensity, tendency, potentiality, disposition) to exert and to feel forces in accordance with certain particular laws (e.g., Coulomb's law).[14] On this view, Lewis's "categorical" properties and DTA's universals are philosophical fictions; they mistakenly drive a "metaphysical wedge" (Ellis and Lierse 1994, 30) between what things are and how they behave. There is nothing

---

14. Nancy Cartwright (1989) and Stephen Mumford (2004) contend that in understanding science, we should invoke natures essentially endowed with causal powers in place of natural laws (rather than invoking essences and powers in an account of what natural laws are). Insofar as a law specifies merely one kind of power (e.g., to exert electric forces) but events (e.g., accelerations) are the net result of powers of many kinds acting in combination, a given law is associated with no regularity among the events it "governs" (since electric forces are always accompanied by gravitational forces, for instance). Accordingly, Cartwright (1983) says, the laws of physics "lie." But why must a regularity account of law say that for every law, there is a corresponding regularity? Cannot a given law merely combine with other laws to entail a nomically necessary regularity? (Recall that perhaps not all nomic necessities are laws.)

categorical (or "structural") that constitutes being electrically charged (or being an electron) and that combines with laws to produce dispositions to behave in various ways. There are just irreducible dispositions grounded in essences.[15]

Essentialism insists that laws support counterfactuals, because laws are metaphysically necessary. But as we have seen, a given accident is preserved under certain counterfactual suppositions despite lacking metaphysical necessity. It is not evident why the law's specific behavior under counterfactual suppositions is a symptom of the law's metaphysical necessity.[16] Indeed it is difficult to see how any of the above approaches to lawhood can account for the specific roles that laws play in scientific reasoning. For example, whatever the precise range of counterfactual suppositions under which the laws are characteristically invariant, it is difficult to see how that range could follow simply from the laws belonging to the "best system," or involving relations of "nomic necessitation," or capturing the causal powers essential to various natural kinds—unless we resort to stipulating the way these facts, in turn, function in counterfactual reasoning. Likewise, although the deductive system with the greatest possible combination of simplicity and strength might as a practical matter be a very useful thing for us to know (in that it packs the maximum information about the world into

---

15. Like Lewis, essentialists regard laws as supervening on the universe's history of property instantiations. But this agreement is misleading: essentialists regard Coulomb's law as immanent in a single instantiation of electric charge, whereas Lewis sees the law as arising from the entire global mosaic. However, some essentialists (see Bigelow et al. 1992; Ellis 2001, 205, 248–252) regard the world itself as having an essence in virtue of which certain laws hold (e.g., global symmetry laws, conservation laws, and laws specifying which natural kinds of particles and fields there are). In that way, every law reflects the essence of *something*, and there is a ground for the counterfactual "If there had been an electron at spatiotemporal location L, then there would still have been protons."

16. That specific behavior—i.e., the laws' collective stability—suggests how the laws can be necessary without being as necessary as metaphysical necessities. Notice also that although essentialism deems laws to be metaphysically necessary, essentialism must avoid deeming all counterlegals to be trivial. Essentialism exploits the intuition that the properties of mass and electric charge, for example, could not retain their identities if they swapped their causal roles (Ellis 1999, 26; 2001, 245; Lierse 1999, 86). Yet some counterlegals have a perfectly respectable place in scientific reasoning. Under essentialism, the counterlegal "Had the gravitational force been twice as strong" must mean roughly "Had there been a force *like* gravity except twice as strong."

the minimum space), it is difficult to see how such a system automatically has the explanatory power or necessity characteristic of laws or why it otherwise should be expected to cut any metaphysical ice, bake any metaphysical bread, or wash any metaphysical windows. Universals and essences, in contrast, may perhaps do some independent metaphysical work. But for them genuinely to account for the natural laws as they figure in scientific practice, the laws' special roles in scientific reasoning cannot simply be built into the notions of "nomic necessitation" or "essence" by hand in an ad hoc fashion.

6. LAWS OF INEXACT SCIENCES?

An "inexact" (or "special") science (such as anatomy, ballistics, ecology, economics, marketing, or psychology) might appear to seek (or perhaps even to have already found) facts that in these sciences play the various roles characteristic of laws. However, there are various obstacles to regarding Boyle's law (that the product of a gas's pressure P and its volume V is constant) as a law of gases, or to regarding Gresham's law (that agents hoard sound money, spending currency of more dubious value) as a law of economics, or to regarding the area law (that larger islands have greater biodiversity) as a law of island biogeography. These obstacles have persuaded some philosophers to deny that inexact sciences have genuine laws of their own.

One obstacle is that any such law would have to come (at least tacitly) with a ceteris paribus qualification (Earman et al. 2002; Earman and Roberts 1999). Though "ceteris paribus" means roughly "all other things being equal," a given qualification may be better captured as the idea that the specified correlation holds "normally," or "in the ideal case," or "in the absence of disturbing factors," or as long as certain other factors have certain values. The law is "hedged" in some way. For example, the gas in a container departs significantly from Boyle's law when its temperature is changed, or some of the gas escapes, or the pressure is high. These circumstances are supposed to be ruled out by the ceteris paribus proviso to Boyle's law. But what exactly does "PV is constant, ceteris paribus" mean?

If it means that PV is constant unless it is not, then the "law" is a trivial, noncontingent truth rather than an interesting discovery. If instead "ceteris paribus" is shorthand for a list of *every* factor allowed by fundamental microphysics and able to cause a gas's PV to vary, then Boyle could not have discovered his law, since he was unaware of some of these factors (e.g., gas molecules adhering to the container's walls or attracting one another). Alternatively, some philosophers, such as Cartwright (1983) and Ronald N. Giere (1988), contend that Boyle's law purports to describe only fictitious "ideal gases" that lack any interfering factors, so Boyle's law has no ceteris paribus qualification and can be used to model actual gases when the relevant purposes tolerate a certain degree of inaccuracy. Likewise, for certain purposes a given system may usefully be approximated as a Mendelian breeding group, and Gregor Mendel's laws are trivially true of this model since they define what a "Mendelian breeding group" is (Beatty 1981; Lloyd 1988; Thompson 1989). Some philosophers regard certain important biological generalizations (such as the principle of natural selection and the Hardy-Weinberg equation) as logically or mathematically necessary yet as able to play the roles characteristic of natural laws (Brandon 1997; Sober 1997; 2000, 124).[17]

In this connection, it is sometimes argued that when laws are used for prediction and explanation, they are applied only on a case-by-case basis and then only with the aid of auxiliary hypotheses or calculational procedures that have been fine-tuned specifically for the given purpose at hand (Cartwright 1983; Morgan and Morrison 1999). Those philosophers who emphasize these features of scientific practice generally accord laws a diminished scientific significance: as mere frameworks or tools for building such models (Cartwright 1999; Giere 1999; Hacking 1983). Science sometimes uses incompatible models to predict or to explain different features of the same system (or even the same feature in different contexts). Science cannot regard all of these models as involving genuine laws.

However, science may not be altogether content to have several radically dissimilar models of the same system (e.g., water as molecular vs.

---

17. Even if the principle of natural selection is logically necessary, it can be a contingent truth that natural selection is a major influence on natural history.

water as a continuous fluid) without a unified explanation of why they all work. That science need not worry about reconciling such models seems like a methodological prescription that science will flout if an attractive reconciliation beckons. Furthermore, scientists seem to use evidence to confirm the applicability of a law or a model, together with various auxiliaries and procedures, to a range of unexamined cases.[18] If a given approach to model building is accurate enough (for certain purposes) regarding all cases of certain sorts, ceteris paribus, then this regularity is contingent and is confirmed inductively, persists under certain counterfactual perturbations, and apparently functions in scientific explanations in the manner of a law—and the problem of "cashing out" the ceteris paribus proviso remains, since that qualification now attaches to a statement of the recipe's range of applicability. Moreover, that Boyle's law purports to describe ideal rather than actual gases seems difficult to reconcile with the fact that Boyle had neither the concept of an "ideal" gas nor an account of what makes a gas "ideal" (e.g., that it consists of molecules without mutual attraction and occupying no finite volume). Such an account is not *part* of Boyle's law. Rather, the extent to which an actual gas has constant PV is *explained* by the extent to which it resembles an ideal gas.

Apparently, then, "ceteris paribus" in Boyle's law refers only to the major disturbing factors (high pressure, changes to the gas's temperature, and so forth), and Boyle was aware of them. There may be no complete list of these factors. Obviously (to shift examples), Gresham's law does not apply if the society is wiped out, or if its members believe that hoarding the sounder currency causes illness, and so forth. As long as this "and so forth" is understood, there is no need ever to cash it out

---

18. For example, computer simulations and model organisms, which obviously cannot *be* laws, are used to predict and to explain and thus are sometimes thought to demonstrate that laws are not central to these tasks in science. But a given model is often intended to be generally applicable, and its range of applicability is confirmed inductively. Of course such an inductive strategy may either mislead (as in Jacques Monod's slogan "What's true for *E. coli* is true for the elephant") or pay off (as when Jan Swammerdam titled his work on muscle contraction "Experiments on the Particular Motion of the Muscles in the Frog, Which May Be Also, in General, Applied to All the Motions of the Muscles in Men and Brutes").

further or to make it more explicit (Wittgenstein 1978, 349, 417). Part of understanding Gresham's law is knowing how to recognize whether some factor qualifies as disturbing. We can catch on to which factors these are without having to read a complete list of them. (A nonexpert may even understand the ceteris paribus proviso without being able herself or himself to tell whether some factor qualifies as disturbing, just as she or he understands other technical terms by virtue of knowing who the relevant experts are to whom she or he should defer.)

However, societal events are ultimately nothing but the outcomes of microphysical processes. Certain sequences of microevents permitted by the fundamental laws of physics involve a society's members hoarding the weaker currency and spending the sounder. In one such sequence, each member of the society happens whenever she or he spends money to forget momentarily which currency is sounder, because (as chance would have it) some neuron in each agent's brain behaves at that moment in a manner that the fundamental microlaws deem extremely unlikely but nevertheless possible. The ceteris paribus proviso to Gresham's law does not rule out this freakish sequence of events, since economists surely do not need to grasp the subtleties of fundamental microphysics to understand the proviso to Gresham's law.

In other words, not all exceptions to a macro-level "law" can be specified in the vocabulary of the macro science (Fodor 1991; Schiffer 1991). For example, it might require physics (or, at least, neurology) to specify a certain circumstance in which an agent would depart from a psychological "law." The ceteris paribus proviso to the law fails to cover those exceptions.

This, then, is a further obstacle to regarding inexact sciences as having genuine laws: the alleged laws are false (Gardiner 1952; Kincaid 1996; Scriven 1959, 1961). Perhaps, however, we should relax the requirement that genuine laws be exceptionless in favor of holding that a law of an inexact science be reliable, that is, sufficiently accurate for the relevant purposes (Mill 1884, 587). The proviso to Boyle's law neglects to mention a host of petty influences that make the PV of actual gases vary somewhat. Yet Boyle's law with its proviso (which rules out the major interfering factors—the ones we cannot afford to neglect) is often close enough to the truth for various purposes in chemistry, theoretical and practical. Understanding Boyle's law requires knowing something about the range of purposes for which it can safely be applied. Likewise, the freakish sequence of neural events that I mentioned earlier is too rare to make Gresham's law unreliable for the purposes of economics.

The limited range of a special science's interests influence which facts qualify as laws of that science. Consider another example: the human aorta carries all of the body's oxygenated blood from the heart to the systemic circulation. This reference to "*the* human aorta", rather than to all or to most human aortas, apparently indicates that we are dealing here with a "generic"—roughly speaking, with a policy of drawing influences defeasibly (i.e., in the absence of further information). This policy, though fallible, is sufficiently reliable for certain purposes (e.g., in medicine for forming expectations about patients in the absence of more specific information regarding them). Whatever the precise truth conditions of a generic, "exceptions" to it obviously do not bar its truth (Achinstein 1965; Carlson and Pelletier 1995; Clark 1965; Molnar 1967).

However, this fact about the human aorta is merely an accident of natural history; it might not have held had evolution taken a different path. Accordingly, some philosophers conclude that such a fact lacks the sort of necessity characteristic of laws (Beatty 1981, 1995; Rosenberg 2001; Waters 1998). On the other hand, medicine does not treat evolutionary history as a variable. A physician might say that the shooting victim would not have survived even if he or she had been brought to the hospital sooner, since the bullet punctured his or her aorta and the human aorta carries all of the body's oxygenated blood from the heart to the systemic circulation. (This fact about the aorta would still have held had the victim been brought to the hospital sooner.) But it would not be medically relevant to point out that the victim might have survived the shooting had evolutionary history taken a different course. Accordingly, perhaps the fact that the human aorta carries all of the body's oxygenated blood to the systemic circulation is a law of human physiology even if it is an accident of physics.

But (to shift examples) even if the "law" that larger islands have greater biodiversity (all other things—such as their distance from the mainland—being equal) is sufficiently reliable for the purposes of island biogeography and even if this "area law" would still have been reliable under some range or other of counterfactual suppositions, what makes it an island-biogeographical *law*? What range of invariance must such a principle exhibit for it to be *necessary* for island-biogeographical purposes? Recall the proposal that a truth *n* is a nomic necessity exactly when *n* belongs to a nontrivially "stable" set, where a nonempty, logically closed set of claims *m* is "stable" exactly when its members are not only true but also would all still have been true under as broad a range of counterfactual suppositions as they could all logically possibly be—namely, under every supposition that is logically consistent with every member. A set's nontrivial stability is associated with its members possessing a variety of necessity. Now what range of invariance must a set's members exhibit for that set to be "stable for the purposes of a given scientific field"? Since the field's concerns are limited, certain claims and counterfactual suppositions lie outside the field's interests, and the field is irrelevant in certain conversational contexts. A "logically closed" set is stable for the purposes of a given science, and hence its members are nomic necessities of that field, if and only if all of its members not only are of interest to the field and reliable for the field's purposes, but also—in every context of interest to the field—would still have been reliable, for the field's purposes, under every counterfactual supposition of interest to the field and consistent with the reliability of all of the set's members. The members of a set that is nontrivially stable for a given field are collectively as resilient under counterfactual suppositions relevant to the field as they collectively could be. Accordingly, they possess nomic necessity for the field.

On this view, an inexact science's laws need not include every detail of the fundamental laws of physics. The features of the laws of physics that are relevant to island biogeography may be captured by various general constraints, such as that a creature cannot get from place to place without traversing the intervening space. Along with the limits of island biogeography's interests, these constraints may suffice (without the fundamental laws of physics) to delimit the range of counterfactual suppositions under which the island-biogeographical laws are stable. Within that range there may even be counterfactual suppositions under which the fundamental laws of physics would not still have held. For example, biological species (considered in an ecological context) would still have been distributed according to the area law even if, contrary to the actual laws of physics, creatures had been made of some continuous rigid substance rather than molecules (cf. Putnam 1975, 296). The factors affecting species dispersal would have been unchanged; for example, smaller islands would still have presented smaller targets to off-course creatures and so would still have accumulated fewer strays, ceteris paribus.

On this view, the laws of an inexact science exhibit a range of stability that in some respects extends beyond the range of stability of the fundamental laws of physics and vice versa. Consequently, the variety of necessity characteristic of the laws of some inexact science is not possessed by the fundamental laws of physics and vice versa. The laws of an inexact science fund explanations that are irreducible in principle to explanations at the level of fundamental physics.

Whether a given inexact science is in fact autonomous is for scientific research to discover. Whether the fundamental laws of physics are privileged among the natural laws (e.g., in having greater generality or in being exceptionless) remains philosophically controversial.[19]

References

Achinstein, P. 1965. "'Defeasible' Problems." *Journal of Philosophy* 62:629–633.
Armstrong, D. 1978. *Universals and Scientific Realism*. Cambridge: Cambridge University Press.
Armstrong, D. 1983. *What Is a Law of Nature?* Cambridge: Cambridge University Press.
Armstrong, D. 1997. *A World of States of Affairs*. Cambridge: Cambridge University Press.
Beatty, J. 1981. "What's Wrong with the Received View of Evolutionary Theory?" In *PSA 1980*, vol. 2, ed. P. Asquith and R. Giere, 397–426. East Lansing, MI: Philosophy of Science Association.
Beatty, J. 1995. "The Evolutionary Contingency Thesis." In *Concepts, Theories, and Rationality in the Biological Sciences*, ed. G. Wolters and J. Lennox with P. McLaughlin, 45–81. Pittsburgh: University of Pittsburgh Press.

---

19. Some of the ideas in this essay are elaborated at greater length in Lange 2000, 2002, 2004, 2005.

Beebee, H. 2000. "The Non-governing Conception of Laws of Nature." *Philosophy and Phenomenological Research* 61:571–594.

Bennett, J. 1984. "Counterfactuals and Temporal Direction." *Philosophical Review* 93:57–91.

Bigelow, J., B. Ellis, and C. Lierse. 1992. "The World as One of a Kind: Natural Necessity and Laws of Nature." *British Journal for the Philosophy of Science* 43:371–388.

Black, M. 1967. "Induction." In *Encyclopedia of Philosophy*, vol. 4, ed. P. Edwards, 169–181. New York: Macmillan.

Braithwaite, R. 1927. "The Idea of Necessary Connection (I)." *Mind* 36:467–477.

Braithwaite, R. 1953. *Scientific Explanation*. Cambridge: Cambridge University Press.

Brandon, R. 1997. "Does Biology Have Laws? The Experimental Evidence." *Philosophy of Science* 64 (supp.): S444–S457.

Carlson, G., and F. Pelletier, eds. 1995. *The Generic Book*. Chicago: University of Chicago Press.

Carroll, J. 1994. *Laws of Nature*. Cambridge: Cambridge University Press.

Cartwright, N. 1983. *How the Laws of Physics Lie*. Oxford: Clarendon.

Cartwright, N. 1989. *Nature's Capacities and Their Measurement*. Oxford: Clarendon.

Cartwright, N. 1999. *The Dappled World: A Study of the Boundaries of Science*. Cambridge: Cambridge University Press.

Chisholm, R. 1946. "The Contrary-to-fact Conditional." *Mind*, n.s., 55:289–307.

Chisholm, R. 1955. "Law Statements and Counterfactual Inference." *Analysis* 15:97–105.

Clark, R. 1965. "On What Is Naturally Necessary." *Journal of Philosophy* 62:613–625.

Davidson, D. 1990. *Essays on Actions and Events*. Oxford: Clarendon.

Dretske, F. 1977. "Laws of Nature." *Philosophy of Science* 44:248–268.

Earman, J., C. Glymour, and S. Mitchell. 2002. *Ceteris Paribus Laws*. Dordrecht, Netherlands: Kluwer Academic.

Earman, J., and J. Roberts. 1999. "Ceteris Paribus, There Is No Problem of Provisos." *Synthese* 118:439–478.

Ellis, B. 1999. "Causal Powers and Laws of Nature." In *Causation and Laws of Nature*, ed. H. Sankey, 19–34. Dordrecht, Netherlands: Kluwer Academic.

Ellis, B. 2001. *Scientific Essentialism*. Cambridge: Cambridge University Press.

Ellis, B., and C. Lierse. 1994. "Dispositional Essentialism." *Australasian Journal of Philosophy* 72:27–45.

Fodor, J. 1974. "Special Sciences (or, The Disunity of Science as a Working Hypothesis)." *Synthese* 28:97–115.

Fodor, J. 1991. "You Can Fool Some of the People All of the Time, Everything Else Being Equal: Hedged Laws and Psychological Explanations." *Mind* 100:19–34.

Gardiner, P. 1952. *The Nature of Historical Explanation*. Oxford: Oxford University Press.
Giere, R. 1988. *Explaining Science: A Cognitive Approach*. Chicago: University of Chicago Press.
Giere, R. 1999. *Science without Laws*. Chicago: University of Chicago Press.
Goodman, N. 1947. "The Problem of Counterfactual Conditionals." *Journal of Philosophy* 44:113–128.
Goodman, N. 1983. *Fact, Fiction, and Forecast*. 4th ed. Cambridge, MA: Harvard University Press.
Haavelmo, T. 1944. "The Probability Approach in Econometrics." *Econometrica* 12 (supp.): 1–115.
Hacking, I. 1983. *Representing and Intervening*. Cambridge: Cambridge University Press.
Hacking, I. 1991. "A Tradition of Natural Kinds." *Philosophical Studies* 61:109–126.
Harré, R., and E. Madden. 1975. *Causal Powers: A Theory of Natural Necessity*. Oxford: Blackwell.
Hempel, C. 1965. *Aspects of Scientific Explanation and Other Essays in the Philosophy of Science*. New York: Free Press.
Horwich, P. 1987. *Asymmetries in Time*. Cambridge, MA: MIT Press.
Huxley, T. H. 1893. *Collected Essays*. 9 vols. London: Macmillan.
Jackson, F. 1977. "A Causal Theory of Counterfactuals." *Australasian Journal of Philosophy* 55:3–21.
Kim, J. 1992. "Multiple Realization and the Metaphysics of Reduction." *Philosophy and Phenomenological Research* 52:1–26.
Kincaid, H. 1996. *Philosophical Foundations of the Social Sciences: Analyzing Controversies in Social Research*. Cambridge: Cambridge University Press.
Kneale, W. 1952. *Probability and Induction*. Oxford: Oxford University Press.
Lange, M. 2000. *Natural Laws in Scientific Practice*. New York: Oxford University Press.
Lange, M. 2002. "Who's Afraid of *Ceteris-Paribus* Laws? or, How I Learned to Stop Worrying and Love Them." *Erkenntnis* 57:407–423.
Lange, M. 2004. "The Autonomy of Functional Biology: A Reply to Rosenberg." *Biology and Philosophy* 19:93–101.
Lange, M. 2005. "Laws and Their Stability." *Synthese* 144:415–432.
Lewis, D. 1973. *Counterfactuals*. Oxford: Blackwell.
Lewis, D. 1979. "Counterfactual Dependence and Time's Arrow," *Noûs* 13 (1979), 455–76.
Lewis, D. 1983. "New Work for a Theory of Universals." *Australasian Journal of Philosophy* 61:343–377.
Lewis, D. 1986. *Philosophical Papers*. Vol. 2. Oxford: Oxford University Press.
Lewis, D. 1999. "Humean Supervenience Debugged." In *Papers in Metaphysics and Epistemology*, 224–247. Cambridge: Cambridge University Press.

Lierse, C. 1999. "Nomic Necessity and Natural States." In *Causation and Laws of Nature*, ed. H. Sankey, 83–88. Dordrecht, Netherlands: Kluwer Academic.

Lloyd, E. 1988. *The Structure and Confirmation of Evolutionary Theory*. New York: Greenwood.

Loewer, B. 1996. "Humean Supervenience." *Philosophical Topics* 24:101–127.

Mackie, J. 1962. "Counterfactuals and Causal Laws." In *Analytical Philosophy*, ed. R. Butler, 66–80. New York: Barnes and Noble.

Mill, J. 1884. *A System of Logic, Ratiocinative and Inductive*. 8th ed. New York: Harper.

Mitchell, S. 2000. "Dimensions of Scientific Law." *Philosophy of Science* 67:242–265.

Molnar, G. 1967. "Defeasible Propositions." *Australasian Journal of Philosophy* 45:185–197.

Moore, G. 1962. *Commonplace Book, 1919–1953*. Ed. C. Lewy. London: Allen and Unwin.

Morgan, M., and M. Morrison, eds. 1999. *Models as Mediators*. Cambridge: Cambridge University Press.

Mumford, S. 2004. *Laws in Nature*. London: Routledge.

Nagel, E. 1961. *The Structure of Science*. New York: Harcourt, Brace, and World.

Peirce, C. 1934. *Collected Papers of Charles Sanders Peirce*. Vol. 5. Ed. C. Hartshorne and P. Weiss. Cambridge, MA: Harvard University Press.

Pollock, J. 1976. *Subjunctive Reasoning*. Dordrecht, Netherlands: Reidel.

Putnam, H. 1975. "Philosophy and Our Mental Life." In *Philosophical Papers*, vol. 2, *Mind, Language, and Reality*, 291–303. Cambridge: Cambridge University Press.

Quine, W. 1960. *Word and Object*. Cambridge, MA: MIT Press.

Reichenbach, H. 1947. *Elements of Symbolic Logic*. New York: Macmillan.

Reichenbach, H. 1954. *Nomological Statements and Admissible Operations*. Amsterdam: North-Holland.

Roberts, J. 1998. "Lewis Carroll and Seeing through the Looking Glass." *Australasian Journal of Philosophy* 76:426–438.

Rosenberg, A. 2001. "How Is Biological Explanation Possible?" *British Journal for the Philosophy of Science* 52:735–760.

Scheffler, I. 1981. *The Anatomy of Inquiry*. Indianapolis: Hackett.

Schiffer, S. 1991. "Ceteris Paribus Laws." *Mind* 100:1–17.

Scriven, M. 1959. "Truisms as the Grounds for Historical Explanations." In *Theories of History*, ed. P. Gardiner, 443–475. New York: Free Press.

Scriven, M. 1961. "The Key Property of Physical Laws—Inaccuracy." In *Current Issues in the Philosophy of Science*, ed. H. Feigl and G. Maxwell, 91–104. New York: Holt, Rinehart, and Winston.

Shoemaker, S. 1980. "Causality and Properties." In *Time and Cause*, ed. P. van Inwagen, 109–135. Dordrecht, Netherlands: Reidel.

Shoemaker, S. 1998. "Causal and Metaphysical Necessity." *Pacific Philosophical Quarterly* 79:59–77.
Sklar, L. 1991. "How Free Are the Initial Conditions?" In *PSA 1990*, vol. 2, ed. A. Fine, M. Forbes, and L. Wessels, 551–564. East Lansing, MI: Philosophy of Science Association.
Sober, E. 1988. "Confirmation and Law-likeness." *Philosophical Review* 97:93–98.
Sober, E. 1997. "Two Outbreaks of Lawlessness in Recent Philosophy of Biology." *Philosophy of Science* 64 (supp.): S458–S467.
Sober, E. 2000. *Philosophy of Biology*. 2nd ed. Boulder, CO: Westview.
Strawson, P. 1952. *Introduction to Logical Theory*. London: Methuen.
Swoyer, C. 1982. "The Nature of Natural Laws." *Australasian Journal of Philosophy* 60:203–223.
Thompson, P. 1989. *The Structure of Biological Theories*. Albany: State University of New York Press.
Tooley, M. 1977. "The Nature of Laws." *Canadian Journal of Philosophy* 7:667–698.
Tooley, M. 1987. *Causation: A Realist Approach*. Oxford: Clarendon.
van Fraassen, B. 1989. *Laws and Symmetry*. Oxford: Oxford University Press.
Waters, C. 1998. "Causal Regularities in the Biological World of Contingent Distributions." *Biology and Philosophy* 13:5–36.
Wigner, E. 1967. *Symmetries and Reflections*. Bloomington: Indiana University Press.
Wittgenstein, L. 1978. *Remarks on the Foundations of Mathematics*. Ed. G. von Wright, R. Rhees, and G. Anscombe. Cambridge: MIT Press.
Woodward, J. 2001. "Law and Explanation in Biology: Invariance Is the Kind of Stability That Matters." *Philosophy of Science* 68:1–20.

Chapter 4

Reading Nature

Realist, Instrumentalist, and Quietist
Interpretations of Scientific Theories

P. KYLE STANFORD

1. PRELIMINARY RECONNAISSANCE: REALISM, INSTRUMENTALISM, AND INTERPRETATION

Questions concerning how we are to interpret our best scientific theories, make sense of what they are telling us, or even just connect them systematically to the world around us have a remarkably long intellectual pedigree.[1] And they have most often been motivated not by the sorts of rarified puzzles we encounter in the course of trying to decide how we could even *possibly* understand claims about, say, superpositions or complementarity in quantum mechanics, but instead by concerns about whether we are really entitled to *believe* what our theories say if we interpret them in what otherwise strikes us as the most literal, natural, or straightforward manner. Before the relatively recent professionalization

---

1. My thanks to Aldo Antonelli; Jeff Barrett; Ludwig Fahrbach; Arthur Fine; Yoichi Ishida; Pen Maddy; David Malament; Mark Newman; Michael Poulin; Jim Weatherall; students in graduate seminars at the University of Pittsburgh Department of History and Philosophy of Science and the University of California, Irvine, Department of Logic and Philosophy of Science; and many others I have inexcusably forgotten for useful discussion and suggestions regarding the material in this chapter.

of academic fields, such concerns were well-represented among the figures who served simultaneously as the leading practitioners and philosophers of science, a fact nicely illustrated by the strident debates throughout this community in the eighteenth and nineteenth centuries concerning whether only pure inductive methods were legitimate for scientific inquiry and/or whether the competing "method of hypothesis" could produce any genuine knowledge of nature (see Laudan 1981, chap. 8).

Having more or less settled that scientific inquiry really does proceed in large part by proposing theories or hypotheses about the entities, events, and dynamical principles at work in otherwise inaccessible domains of nature, the position known as scientific realism has often seemed the most natural or straightforward approach to interpreting those hypotheses. The scientific realist is a forthright and commonsensical soul who tells us that our best scientific theories simply offer (at least probably and/or approximately) true descriptions of how things stand in the world, including most contentiously those domains or aspects of the world about which it is extremely difficult to get information in any other way because the events and entities in them are extremely small, unavoidably remote, occurred in the distant past, or are otherwise inconveniently situated. But the realist also typically holds a distinctive position concerning how the truth of those theories is to be understood. Her "correspondence theory" of truth holds that the claims of scientific theories or hypotheses are true when the propositions they express "correspond to" or mirror the way the world or the natural order really is. Thus, she suggests, the claims of our best physical chemistry about negatively charged electrons orbiting atomic nuclei should be understood to be just like claims about the cherries in a cherry pie, although they are admittedly much more difficult to confirm or refute.

In the hands of thinkers like Richard Boyd and the early Hilary Putnam, belief in such correspondence truth for at least our best scientific theories was touted as part of the only naturalistic or scientific explanation of the success of the scientific enterprise itself. More recently, Philip Kitcher (2001) has emphasized the fundamental continuity of this realist line of thinking about scientific theories with inferences about success, truth, and correspondence to the world that we draw in more familiar and everyday contexts. Kitcher's "Galilean strategy" argues that whenever the guides to the world (e.g., subway maps) we are using allow us to attain the kind of fine-grained and systematic success provided by many of our best contemporary scientific theories, we routinely and unproblematically infer that they do so because they reflect, or mirror, or correspond to the way the world actually is: that the order of the subway stations in the world really does follow the order in which they appear on the map, for instance. Kitcher points out that we have no trouble making sense of the correspondence metaphor in such contexts. And he argues that we have no compelling specific reason to doubt the effectiveness of this general "success-to-truth" inference in the case of theoretical science, just as Galileo's opponents could offer no compelling specific reason to doubt that his telescope was any less reliable and effective when pointed toward the heavens than he and others had demonstrated it to be in innumerable independently verifiable terrestrial applications like reading the writing on distant buildings or identifying the people on far-off ships coming into port.

The scientific realist, then, holds a distinctive combination of views about our best scientific theories. Such theories are to be understood as offering straightforward descriptions that correspond to how things stand in the world itself, and so understood, we are (at least typically) justified in believing what they say.[2] On this account, scientific theories are like fascinating travelogues about distant and exotic places: a realist

---

2. We might instead describe scientific realism in a broader fashion that makes room for a deflationary conception of truth as well. That is, one might suggest that the realist is really committed simply to the following conjunction of views: that the claims of theoretical science should be interpreted in just the same way that more everyday claims about tables, chairs, and cats on mats are and that (so interpreted) we have sufficient reason to believe them. Thus, although deflationism about truth seems more naturally mated with quietism about the interpretive status of scientific theories (see below), it may be this broad interpretation of realism which invites the accusation that quietists are just realists in disguise. It is ironic, then, that because so-called internal realists *explicitly* abandon the correspondence conception of truth in favor of a coherentist or pragmatist alternative (while maintaining that we have good reasons to believe the claims of scientific theories to be true under such a pragmatist conception), such internal realism is widely regarded as not really being a form of scientific realism at all. Of course, the labels are unimportant so long as we are clear both on what epistemic attitude we think we are justified in taking toward the claims to truth of our successful scientific theories and under what conception of their truth this attitude is warranted.

will interpret the claim that electrons carry a negative charge or that we share a distant common ancestor with spiders and elephants simply as telling us how things stand in parts or aspects of the world about which it is otherwise quite difficult to get information. In just this same way, a realist will interpret the quantum mechanical claim that particles in a given experimental setup exist in a superposition until a measurement collapse forces them to assume determinate spatial positions as a straightforward report of how things stand in nature itself. Variants of the realist position also come in such flavors as the "structural" realism of John Worrall and the "entity" realism independently defended by Ian Hacking and Nancy Cartwright, which advocate taking this realist attitude only toward particular aspects or parts of our best scientific theories, such as the structure they ascribe to nature or the entities they recognize.

The realist stands most directly opposed by the members of a long-standing minority tradition I will simply call "instrumentalism." The instrumentalist doubts whether simply reporting the facts about otherwise inaccessible domains or aspects of nature is what even our most successful scientific theories actually do. But such doubt can originate in more than one way, and the grounds for the instrumentalist's caution are as diverse as the challenges facing the realist position itself. These range from the relatively narrow appeal of a phenomenalist conception of the world or a verificationist conception of the meanings of our terms quite generally to the comparatively wider influence of the so-called pessimistic induction over the history of science and the underdetermination of theories by evidence. The pessimistic induction points out that even the most successful scientific theories of the past have ultimately proved to be mistaken, and it suggests that we have no convincing reason to expect the ultimate fates of our own theories (which are successful in just the same ways, even if sometimes to a greater extent or degree) to be any different. The problem of underdetermination suggests instead that there is always more than one available theory capable of accommodating any finite collection of empirical data and thus that we are never in a position to confirm just a single theory over its competition. More recently, I have argued that the most important challenge for scientific realism arises from what I call the problem of unconceived alternatives

(Stanford 2006). I suggest that the historical record reveals our repeated failure to even *conceive* of scientifically serious alternatives to the best scientific theory concerning a given natural domain proposed at a given time even when such alternatives were also well confirmed on the whole by the available evidence and thus that we have every reason to believe that similarly serious, promising, and well-confirmed alternatives to our own scientific theories presently exist unconceived by us.

This provides some sense of the range of considerations that have motivated instrumentalists to explore general alternatives to the realist conception of scientific theories. Of course resistance can also arise from more local sources, such as the suggestion that we are simply unable to make any literal sense out of the sorts of claims about superpositions and measurement collapses given above and thus that they are not even candidates for "corresponding with reality." But such local grounds for resisting realism serve simply to motivate interest in applying (or reinventing) in the case of some particular scientific theory one or another alternative approach to the interpretation of scientific theories that instrumentalists have sought to develop more generally.

One traditional strand of such instrumentalist thinking has challenged whether scientific theories should be understood as making the claims that the scientific realist sees them as making in the first place. Such semantic instrumentalists suggest either that truth itself should not be conceived of in the realist's preferred manner (whether for the claims of theoretical science or quite generally) or that scientific claims require some special semantic understanding or analysis. The other influential strand of instrumentalist thinking has concentrated its dissent instead on the realist's view of our epistemic entitlements: it does not fault the realist interpretation of the claims of our scientific theories but argues instead that we are not justified in actually believing those claims to be true. Such epistemic instrumentalists have sought to articulate an alternative cognitive attitude toward successful scientific theories to which they think we can indeed be epistemically entitled. What unifies these two strands of instrumentalist thinking is their common commitment to an alternative positive construal of scientific theories as mere conceptual tools or "instruments" for achieving our practical goals and objectives rather than descriptions of nature that are either probably or approximately true in the realist's correspondence sense. It is noteworthy that the phrase "the instrumentalist interpretation of scientific theories" is sometimes used to refer to just one of these alternatives, is sometimes ambiguous between them, and is sometimes used to refer to them jointly.

Of course, within the semantic strand of instrumentalism there is an important distinction between those who think that a special analysis or understanding of truth is only required for scientific claims (perhaps along with some other specific parts of our language) and those who think that some form of instrumentalist conception of truth is the right analysis for any and all claims whatsoever. The latter group is the inheritor of the intellectual legacy of American pragmatism (it was, after all, John Dewey who coined the term "instrumentalism"), with all of the attractions and challenges involved in its embrace of a distinctively pragmatic conception of truth: essentially that the truth about the world just is whatever beliefs about it we will (or idealized agents would) come to accept in the limit of inquiry (C. S. Peirce) or the beliefs about the world that experience would never encourage us to give up were we to adopt them now (William James). In the philosophy of science this position is probably best captured by the "internal realism" variously defended by (a later version of) Putnam and by Brian Ellis. But this version of the instrumentalist position is perhaps the least widely discussed among philosophers of science, in large part because it allows that the truth of scientific claims (though perhaps "merely instrumental" in the requisite sense) is nonetheless perfectly on a par with the truth of all of our other kinds of claims about the world. Thus there is no *special* challenge or concern here about the truth of scientific theories or claims: the further philosophical work to be done concerns the nature of truth in general and thus neither raises issues of special significance to nor engages the particular expertise of most philosophers of science.

The quite different view that the claims of theoretical science require a distinctive kind of semantic reinterpretation or translation not shared by all claims whatsoever was perhaps most influentially pursued in the modern era by a number of the logical positivists, including Percy Bridgman and the early Rudolph Carnap. Initially, this took the form of arguing that the very meaning of the claims of theoretical science is exhausted by whatever implications they have concerning observations or observation statements. But the determined efforts of these thinkers (most notably Carnap) to effect a general reduction of scientific language to such a privileged phenomenological or observational basis ultimately foundered, and this reductionist project was eventually given up for lost even by some of its most important original architects (including Carnap himself). Some later positivists and logical empiricists would preserve the spirit of this broadly semantic approach, however, while giving up these reductionistic ambitions and adopting instead one of two influential views about the semantics of the claims of theoretical science. According to the first, such claims are devoid of *any* semantic content beyond the license they provide to draw inferences between observable states of affairs: they are not assertions about the world at all (and therefore do not have truth conditions) but are simply "inference tickets" from one observable state of affairs to another. On the second view, scientific language making putative reference to theoretical or unobservable entities or states of affairs is both genuinely assertoric and irreducible but nonetheless can in principle simply be eliminated from any given scientific theory altogether, producing a functionally equivalent linguistic apparatus that makes claims only about observable matters of fact. William Craig formulated and proved an influential theorem widely taken to show that such a transformation could in principle be made for any restricted part of the vocabulary of any (recursively axiomatized first-order) theory that was mutually exclusive of and mutually exhaustive with the rest of the (nonlogical) language of that theory. A fortiori, as Carl Hempel pointed out in connection with his famous "theoretician's dilemma," this could be done for a (mutually exclusive and exhaustive) partition of a theory's language into "theoretical" and "observational" components, guaranteeing the eliminability of whatever part of our scientific description of some part of the world traffics in unobservable entities, events, and processes while leaving the theory's observational consequences unchanged.

Even at the time, however, this claim of in principle eliminability was criticized as being of little genuine significance. Ernst Nagel (1961, 136–137) pointed out, for instance, that the axioms of the transformed theory would be infinite in number and correspond one-to-one with all the true statements expressible using the restricted observational vocabulary (rendering them "quite valueless for the purposes of scientific inquiry") and that the transformation could only proceed after one knows (in advance of any deductions from them) all the true statements in the restricted observational language. In the intervening decades philosophers of science have only found more reasons to resist the philosophical as well as the practical significance of such Craig's theorem–style eliminability. Perhaps most important here are the profound differences that have emerged between the sorts of artificial formal systems to which tools like Craig's theorem apply and real scientific theories, which turn out not to admit of any neat division into strictly theoretical and observational vocabulary after all and often not to be especially helpfully understood as axiomatic formal systems in any case. Likewise, philosophers of science have found little to encourage them to adopt the view that the claims of theoretical science are not genuinely assertoric and many reasons to be skeptical of this counterintuitive analysis of the semantics of these apparently descriptive and truth-valued claims. Neither the "inference-ticket" view of scientific theories nor the claim of in principle eliminability has turned out to have much to recommend it as an approach to interpreting or understanding scientific claims beyond the hope of making the instrumentalist position defensible, and this now strikes even most instrumentalists as too high a price to pay for the privilege.

Perhaps understandably, then, in the decades since the heyday of logical empiricism philosophers of science have become increasingly and systematically skeptical of the idea that instrumentalism should be implemented by way of a special understanding of the truth, meaning, or semantic status of the claims of theoretical science. But the declining fortunes of this first, semantic approach to instrumentalism have been accompanied by a substantial revival of interest in the second, epistemic approach. This epistemic strand of instrumentalist thinking does not construe the truth or meaning of scientific claims in a special way but instead simply denies that we have sufficient justification for believing that even the central claims of even our best scientific theories are in fact true.[3] Instrumentalists of this

---

3. Note that such an instrumentalist need not accept the realist's correspondence conception of truth. She might simply allow, for example, that the semantics and truth

distinctively epistemic variety have therefore sought to identify some cognitive attitude we might take toward the claims of our scientific theories besides simply believing the descriptions they seem to offer of otherwise inaccessible domains of nature. Some influential recent versions of this approach have recommended that we believe only the implications our theories have concerning observable matters of fact (Bas van Fraassen's constructive empiricism) or believe only that our theories are effective tools for solving the various empirical and theoretical problems we face (Larry Laudan's problem-solving model of scientific progress).

2. LOSING SIGHT OF A DISTINCTION

Among the problems facing any attempt to formulate and defend instrumentalism, perhaps most fundamental and interesting of all is a deep and persistent line of argument to the effect that the intended contrast between interpreting scientific theories literally, realistically, or at face value, on the one hand, and as mere instruments, on the other, cannot ultimately be made out in any coherent way. Nagel famously posed a version of this challenge against the distinctively semantic variety of instrumentalism popular among his contemporaries. He argued that there was a "merely verbal difference" between the instrumentalist's contention that a given theory offers satisfactory techniques of inference or a reliable "inference ticket" from some observable states to others and the realist's contention that the theory is simply true (Nagel 1961, 139). But much the same challenge continues to be posed to the more recently influential and distinctively epistemic varieties of instrumentalism that instead question

> conditions for the claims of theoretical science are on a par with those of more familiar claims, whether the latter are understood in terms of a pragmatic, deflationary, correspondence, or some other conception of truth. Of course this creates room to endorse a distinctive semantics for theoretical (or other scientific) claims *as well as* the view that we ought not believe the central claims of even our best scientific theories even when they are so understood. But this is surely as it should be. Notice, for example, that an internal realist or pragmatic conception of truth will not draw the sting from the pessimistic induction, which concerns the *transient* character of even the most successful past scientific theories, no matter how their truth is understood.

whether we are justified in believing the descriptions of the world offered by our scientific theories to be true. Paul Horwich (1991), for example, points out that some accounts of belief simply identify it as the mental state responsible for use and suggests that epistemic instrumentalists will have to show why such accounts are mistaken before they are entitled to conclude that their position is any different from that of their putative realist opponents. Howard Stein (1989) offers a much more detailed argument to the effect that, once both realism and epistemic instrumentalism are sophisticated in ways that are independently required in any case, there remains no important difference between them. Realism must abandon, he argues, any pretensions to metaphysically transcendent theorizing or noumenal truth and reference and must also give up the idea that any property of a theory might somehow explain its success in a way that does not simply point out the use that has been made of the theory itself. Instrumentalism, by contrast, must enrich the conception it holds of the functioning of a theory *as an instrument* to include not only calculating experimental outcomes but also adequately representing phenomena in detail across the entire domain of nature to which it can be fruitfully applied and (perhaps most importantly of all) serving as the foundation, resource, and starting point for all further inquiry into that domain. Once the realist's ambitions for our scientific theories have been appropriately restricted and the instrumentalist's ambitions for them appropriately expanded in this way, Stein argues, there is simply no difference that makes a difference left between the two positions, and he suggests that the twin aspects of the resulting view of theories have always been present simultaneously in the deepest scientific work in any case.[4]

The suggestion that there is no room for a distinction between realist and instrumentalist attitudes toward scientific theories has been developed in perhaps the greatest detail, however, by Simon Blackburn (1984, 2002). Blackburn's representative instrumentalist (van Fraassen) urges that we not believe scientific theories but instead merely "accept" them as "empirically adequate"; that is, we are to believe what they say about observable phenomena while remaining agnostic about their further

---

4. This last suggestion is a close relative of one defended in much greater detail by Larry Sklar (2000).

claims concerning unobservables. But as Blackburn (2002, 117–119) notes, the character of the acceptance van Fraassen recommends rightly involves a great deal more than simply using the theory to calculate predictions concerning observables:

> The constructive empiricist is of course entirely in favor of scientific theorising. It is the essential method of reducing phenomena to order, producing fertile models, and doing all the things that science does. So we are counselled to *immerse* ourselves in successful theory.... Immersion will include acceptance as empirically adequate, but it includes other things as well. In particular it includes having one's dispositions and strategies of exploration, one's space of what it is easy to foresee and what difficult, all shaped by the concepts of the theory. It is learning to speak the theory as a native language, and using it to structure one's perceptions and expectations. It is the possession of habits of entry into the theoretical vocabulary, of manipulation of its sentences in making inferences, and of exiting to empirical prediction and control. Van Fraassen is quite explicit that all of this is absolutely legitimate, and indeed that the enormous empirical adequacy of science is an excellent argument for learning its language like a native.... Immersion, then, is belief in empirical adequacy plus what we can call being "functionally organized" in terms of a theory.

Like Stein, Blackburn here insists that instrumentalists recognize the full range of useful instrumental functions that our scientific theories perform. And once this recognition is complete, he suggests, there remains no room for a distinction between the thoroughgoing "immersion" in or "animation" by our theories that instrumentalism recommends and realism itself:

> The problem is that there is simply no difference between, for example, on the one hand being animated by the kinetic theory of gases, confidently expecting events to fall out in the light of its predictions, using it as a point of reference in predicting and controlling the future, and on the other hand believing that gases are composed of moving molecules. There is no difference between being animated by a theory according to which there once existed

living trilobites and believing that there once existed living trilobites.... What can we do but disdain the fake modesty: "I don't really believe in trilobites; it is just that I structure all my thoughts about the fossil record by accepting that they existed"? (Blackburn 2002, 127–128)

These passages frame the problem in perhaps its starkest form. As the instrumentalist strengthens "commitment," "immersion," or "acceptance" to capture all of the various respects in which we rely on our theories instrumentally, she seems to simultaneously eradicate any distinction between such "merely" instrumental reliance and belief itself. And she thus winds up hard-pressed to say what could distinguish a commitment to the *robust* instrumental utility of our best scientific theories from simply believing those same theories to be true.[5]

According to Blackburn, it is no accident that we make the mistake of thinking that there must be a well-posed distinction between realism and instrumentalism, for there are genuine differences in the character or degree of acceptance of a theory in actual scientific contexts to which the supposed divide between realists and instrumentalists might seem closely related. But he argues that closer scrutiny reveals these apparent similarities to be specious.

---

5. One note of caution is in order here. Van Fraassen sometimes seems to suggest that it is only practicing *scientists* who must "immerse" themselves in the theories they work with, while philosophical interpreters of scientific activity (presumably including scientists when they act in this role) are free to rest content with mere belief in the empirical adequacy of those same theories. Whatever merits such a two-tiered view of epistemic commitment might or might not have, Blackburn's challenge retains its bite, for van Fraassen (1980, 73) argues that his constructive empiricism "makes better sense of science, and of scientific activity" than the realist alternative, and this claim is rendered suspect if the "immersion" he is forced to recommend to scientists is itself indistinguishable from belief (i.e., requires from practicing scientists self-delusion on a massive scale). Perhaps partly for this reason, van Fraassen (1980, 81) argues explicitly that even the working scientist's "immersion in the theoretical world-picture does not preclude 'bracketing' its ontological implications." Furthermore (and independently), many of the features of immersion to which Blackburn and Stein draw our attention (e.g., the propriety of using the theory as the foundation for our future inquiry in a given scientific domain) are those about which the philosophical interpreter of science cannot afford to remain agnostic in any case.

It is natural, for instance, to contrast the empirical adequacy of a theory to date with its final and total empirical adequacy. This contrast is itself perfectly legitimate, but it cannot mark any part of the difference between van Fraassen's instrumentalist and her realist opponent, Blackburn suggests, because the fully immersed instrumentalist is no less committed to a theory's continuing adequacy into the future (i.e., the *total* empirical adequacy of the theory) than is the realist. After all, a commitment to the empirical adequacy of a theory is a commitment to the truth of what it says about observable phenomena, full stop, whether in the past, present, or future and whether actually observed or not. Of course in some particular case the instrumentalist might have special reasons for doubting that a theory that has been empirically adequate to date or over some restricted range will continue to be so, but this is a difference of epistemic commitment drawn from *within* the instrumentalist position and thus cannot constitute the difference between such instrumentalism and the realist alternative. And Blackburn's point here would seem to apply equally well to Laudan's formulation of the instrumentalist commitment to the *ongoing* or *continued* ability of a theory to solve empirical and theoretical problems.

In a similar vein, Blackburn notes that we can legitimately contrast full and unreserved acceptance of a theory with embracing it in a more cautious or tentative spirit but argues that this distinction is also inapposite: van Fraassen's fully immersed instrumentalist embraces the empirical adequacy of a scientific theory with no less confidence or conviction and in no more tentative a spirit than the realist embraces the truth of that same theory. As before, there may be room for greater and lesser degrees of confidence or commitment with regard to particular theories, but once again this potential variation will have to be recognized from *within* the instrumentalist position in a manner precisely parallel to the varying degrees of confidence a realist might have in the truth of a theory, and so it cannot mark the distinction between the two positions. And again the same point applies in just the same way to the Laudanian instrumentalist's confidence in a theory's ability to solve our empirical and conceptual problems.

I will here set aside what Blackburn cites as a third possible source of spurious plausibility for the distinction between realism and instrumentalism: the contrast between believing a theory and accepting that theory "in an instrumentalist frame of mind" (Blackburn 2002, 121) while "bracketing it's ontological implications" (van Fraassen 1980, 81; quoted in Blackburn 2002, 117) because we see the theory as violating norms of consistency or appropriate description (quantum mechanics is perhaps the paradigmatic example here). He notes that this cannot be the distinction van Fraassen has in mind, because the latter's constructive empiricist does not limit his recommendation of mere acceptance rather than belief to theories facing special challenges of this sort (and again, we might note, the same is true for Laudan). I suspect that Blackburn does his own argument a disservice here, for the central question he raised was whether the distinction between belief and mere acceptance could be made out at all, not where we could legitimately invoke it. If he really means to recognize a difference between genuine belief and such acceptance with ontological "bracketing" or in an "instrumentalist frame of mind" in these cases, this looks like a plausible candidate distinction between realism and instrumentalism quite generally: the remaining issue would not be whether there was a distinction van Fraassen could make use of but whether he had sufficient reason to consign so many theories to the instrumentalist side of it. (Notice that, unlike the other two distinctions he considers, Blackburn does not show how this distinction is itself one that van Fraassen will need to be able to draw as a difference in degree of commitment to a theory from *within* the instrumentalist position in any case.) To put the matter another way, although Blackburn is right to say that the distinction between theories that face special problems of norm violation and those that do not cannot be the distinction van Fraassen has in mind, it simply does not follow that the distinction between belief and acceptance in an instrumentalist frame of mind (while bracketing ontological implications), if recognized as genuine, cannot be the distinction van Fraassen has in mind.

However this last issue may stand, once any spurious sources of intuitive plausibility for the distinction between realism and instrumentalism are cleared away, the instrumentalist is left with a vexing problem. She seems to owe us some account of how an instrumentalist attitude toward our theories that is powerful and comprehensive enough to capture and support our ordinary scientific practices will not be simply indistinguishable from belief in those same theories. And as we will see in the next section, we will be forced to decide whether we think there is any genuine distinction between the realist and instrumentalist positions even if we ultimately decide to reject *both* in favor of a recently influential alternative that we might describe simply as "quietism" about the interpretation of scientific theories. We turn now to an exploration of this quietist position, which seeks to move beyond both realism and the various forms of instrumentalism by giving up their shared sense that scientific claims and theories stand in need of any kind of "interpretation" or further philosophical analysis at all.

3. KEEPING QUIET

The quietist insists not only that our scientific theories do not need to be "interpreted" in the first place but also that the very enterprise of trying to do so is misguided: like most of the claims we make about the world, she tells us, scientific theories mean just what they say, no more and no less, with no need for a special metatheoretical interpretive stance or account that translates them for us or tells us in more philosophically rigorous terms what they "really" mean or amount to at the end of the day. This position is represented most influentially by Arthur Fine's Natural Ontological Attitude (NOA, pronounced "Noah") but also and more recently by Blackburn (2002) as well. Fine (1986a, 128) begins by identifying a set of commitments that he suggests both the realist and the instrumentalist (called "the antirealist" in the quotation below) hold in common:

> It seems to me that both the realist and the antirealist must toe what I have been calling 'the homely line.' That is, they must both accept the certified results of science as on par with more homely and familiar supported claims.... Let us say, then, that both realist and antirealist accept the results of scientific investigations as 'true,' on par with more homely truths.... And call this acceptance of scientific truths the 'core position.' What distinguishes realists from antirealists, then, is what they add onto this core position. (footnote omitted)

That is, both realists and instrumentalists accept the truths that emerge from scientific inquiry as well established (just like more familiar claims), but realists do this by way of their correspondence conception of truth, while instrumentalists do so either by way of reconceiving what such truth amounts to (whether in general or just for scientific claims) or in virtue of their willingness to "accept" theories, or "use" them instrumentally, or treat them as "empirically adequate" in a perfectly thoroughgoing way while hastening to remind us that they do not *really* believe what they take those theories to say. After embracing this core position, both realists and instrumentalists feel the need to *supplement* it in their various distinctive ways: either with a foot-stamping shout of "really," or with a special analysis of what the truth of scientific claims consists in, or with the admonition that we are simply not to *understand* our (merely instrumental though complete) acceptance of scientific truths in the same way that we understand our acceptance of the more homely truths of our everyday lives. By contrast, Fine suggests that the core position itself is all that we really need, and he recommends that we simply embrace it without any of these further, extrascientific commitments. In the process we should dispense with the idea that scientific theories themselves stand in need of any special philosophical interpretation or analysis: "What binds realism and antirealism together," he says, is that "[t]hey see science as a set of practices in need of an interpretation, and they see themselves as providing just the right interpretation. But science is not needy in this way" (Fine 1986a, 147–148).

As even the name he gives to his own position suggests, Fine denies that scientific realism is somehow the obvious default or "natural" attitude to take toward the claims of our scientific theories. Only a perverse felt need to provide scientific inquiry with some kind of external authentication or grounding, he suggests, would lead us to append to it a special philosophical theory of what such truth amounts to and declare this to be its most natural "interpretation." He argues that it is surely much *more* natural simply to recognize that science gives us claims that are true in the homely way that NOA allows above but not insist on any further specification or analysis or interpretation of what such truth really amounts to in the end, just as (he suggests) we decline to do with the more familiar claims we rely on in the course of our ordinary lives. Thus,

Fine (1986b, 177) suggests, the "naturalness" of NOA is "the 'California natural'—no additives, please!"

Indeed, a crucial aspect of Fine's view is its stubborn refusal to provide or ratify any general account of what "truth" is. He tells us that NOA instead "holds a 'no-theory' conception of truth" (Fine 1986b, 175) in large part because it denies that truth is any one thing: "Of course we are all committed to there being some kind of truth. But need we take that to be something like a 'natural' kind? This essentialist idea is what makes the cycle [from realism to instrumentalism and back again] run, and we can stop it if we stop conceiving of truth as a substantial something—something for which theories, accounts, or even pictures are appropriate" (Fine 1986a, 142). While NOA accepts our ordinary uses of the term "true," it denies "that truth is an explanatory concept, or that there is some general thing that makes truths true" (Fine 1986b, 175). Fine applies the same antiessentialism to the very idea that the processes or products of scientific inquiry have a general aim or goal or (therefore) admit of any general interpretation. Of course we certainly recognize goals or aims or purposes in scientific contexts, but Fine insists that these are local in character and internal to the practice of science itself: "For what purpose is this particular instrument being used, or why use a tungsten filament here rather than a copper one?" It is simply a gross fallacy in quantifier logic, he notes, to move from "They all have aims" to "There is an aim they all have" (Fine 1986b, 173). And when we ask after the goal, or aim, or purpose of science itself, "we find ourselves in a quandary, just as we do when asked 'What is the purpose of life?' or indeed the corresponding sort of question for any sufficiently rich and varied practice or institution" (Fine 1986a, 148). "The quest for a general aim [for science], like the quest for the meaning of life, is just hermeneuticism run amok" (Fine 1986b, 174). NOA invites us instead to see science as standing on its own bottom and emphatically *not* standing in need of any overarching external goal or aim, whether realist or instrumentalist in character. Rather, "NOA suggests that the legitimate features of these additions are already contained in the presumed equal status of everyday truths with scientific ones, and in our accepting them both as *truths*. No other additions are legitimate, and none are required" (Fine 1986a, 133).

In a widely influential discussion, Alan Musgrave (1989) has suggested that Fine does not actually manage to navigate between the realist and instrumentalist commitments he seeks to avoid precisely *because* the "core position" Fine claims for NOA as the common foundation for realism and instrumentalism accepts "the results of scientific investigations as 'true,' on par with more homely truths" (Fine 1986a, 128) and the "equal status of everyday truths with scientific ones" (Fine 1986a, 133). Musgrave insists that Fine must either understand such truth in the realist's preferred way, in which case he and NOA become realists after all (the outcome he prefers), or fail to articulate any coherent view (or even a coherent "attitude"?) at all.

The dilemma is sharpened, Musgrave argues, by considering the few things NOA does say about truth. First, Fine (1986a, 130) insists that NOA treats "truth in the usual referential way, so that a sentence (or statement) is true just in case the entities referred to stand in the referred-to relations" and "sanctions ordinary referential semantics, and commits us, via truth, to the existence of the individuals, properties, relations, processes, and so forth referred to by the scientific statements that we accept as true." In a similar spirit Fine (1986a, 133) goes on to explain that NOA "recognizes in 'truth' a concept already in use and agrees to abide by the standard rules of usage. These rules involve a Davidsonian-Tarskian referential semantics, and they support a thoroughly classical logic of inference. Thus NOA respects the customary 'grammar' of 'truth' (and its cognates)." So NOA explicitly allows that the statement "There is a full moon tonight" is true if and only if there is indeed a full moon tonight (Musgrave's example 1989, 388). But compare this with a claim like the following: "The statement 'George W. Bush gives me the creeps' is true if and only if George W. Bush gives me the creeps" (Musgrave's example, updated). We endorse the ordinary referential semantics and rules of usage here only because we think the statement (even as it appears without internal quotation marks, on the right hand side) should be understood or interpreted as an idiom that is *not* to be taken (as Musgrave says) "at face value": rather than recognizing the existence of an entity or a group of objects ("the creeps") that are routinely presented to the speaker by Bush, we interpret the statement as saying something like Bush makes the speaker nervous and/or uncomfortable.

What Musgrave argues is that NOA's commitment to the idea that scientific claims are true in just the way that more homely truths are must mean *either* that it uses ordinary referential semantics *and* accepts the truth of scientific claims when they are taken at "face value" (as we take the claim about the full moon)[6] *or* that it really does take *no position* on whether scientific claims are to be taken at face value rather than understood in some idiomatic or nonliteral way (like the creeps that Bush gives me). In the former case, instrumentalists do not actually accept the "core position" Fine describes, because they deny that we are justified in holding scientific claims to be true at face value in Musgrave's sense (either because scientific claims should not be accepted as true at all or because they should be reinterpreted before being accepted as true) and by endorsing the "core position" alone, Fine and NOA simply turn out to be realists after all. If instead NOA really means to remain agnostic concerning whether or not scientific claims should be taken at face value in this way, Musgrave insists that there is no *one* core position common to realism and the various forms of instrumentalism but rather many different core positions depending on what sense we give to the view that scientific claims are "true" in just the same way that the homely truths of our everyday lives are. These different positions enjoy only the *apparent* unity that they can all agree to a common verbal formula, just as Berkeley could still use the same *expressions* that the man in the street would use to say things about the world, even though he meant by them something far removed from what the man in the street meant (or thought he did, in

---

6. Of course this sense of "face value" must be quite different from the homely acceptance as true that Fine uses to describe the core position, above, for the latter was explicitly intended *not* to prejudge the character of our acceptance at all. Musgrave's sense of taking a statement at "face value" might be best described as treating the statement's fundamental ontological commitments (to entities, events, properties, relations, etc.) in precisely the manner suggested by its surface grammar. Of course this particular example is a curious one even for Musgrave's sense of "face value," as we do not really think the moon is "full" of anything but rather that a large proportion of its visible surface is illuminated. (Indeed I suspect that this sort of consideration should engender some suspicion about the idea that metaphorical or idiomatic uses of language can be clearly or cleanly separated from literal ones in general, but I will not pursue the matter here.) In any case, I suggest below that Fine would deny that Musgrave's sense of "face value" was ever the matter at issue.

any case) when he used those expressions. And if the "core position" that exhausts NOA is based on this sort of simple equivocation, Musgrave argues, then NOA is really just a philosophical know-nothingism: what the NOAer is really saying, with an infuriatingly genial smile, is something like "I am happy to say that electrons orbit atomic nuclei, but I don't really mean anything in particular by that." Moreover, as NOA's "core position" holds that scientific claims are true in just the same way that more homely truths are, it proudly insists on knowing nothing not just about science but about everything else as well.

Of course something seems profoundly misguided about Musgrave's analysis—it does not take sufficiently seriously the fact that NOA thinks there is nothing useful to be said in general about how "truth" should be understood. To be sure, there are uses of language whose accepted meaning does not follow their surface grammar, but Fine's point was that there is nothing (that is, no *one* thing) to say about the truth of claims in general (or even just the truth of scientific claims in general) even *after* they are all "normalized" into a form that can be taken at what Musgrave calls "face value" (e.g., from "George W. Bush gives me the creeps" to "George W. Bush makes me nervous and/or uncomfortable"). NOA insists that truth no more admits of a unitary analysis after such transformations (and for claims that do not need them in the first place) than before. The various things that the truth of various claims might amount to, in science as in everyday living, are multiple and many-splendored: we are guided in what to make of our acceptance (or the "truth") of a particular scientific claim in a particular context by circumstances specific to it, *and that is all there is to say.*[7] In particular there is simply nothing more that is both useful and general to say about truth

---

7. Musgrave endorses Fine's skepticism about the idea that there is some one thing that makes all truths true, but he nonetheless seems to want to treat scientific claims as an identifiable *category* whose interpretation and/or truth conditions can be characterized in a uniform way (e.g., like claims about "the moon" or like claims about "the creeps"). But this is to seriously underestimate the sort of heterogeneity about truth that Fine means NOA to recognize. It is not clear to me that this refusal to see the point is not intentional and studied, however, as Musgrave does sometimes seem to be responding to NOA in the same ironic spirit in which Samuel Johnson or G. E. Moore responded to skepticism, in this case by pointing out just what he thinks can indeed be said about the truth of such claims that is both general and useful in this connection.

itself or about the kind(s) of truth that broad categories of claims (like the claims of science or the claims of everyday living) enjoy. Likewise, Musgrave's taunt of know-nothingism is a little unfair, for NOA insists on knowing nothing at just the point where it also insists that there is *nothing more to know*, at least nothing of the global, philosophical sort Musgrave has in mind.

Of course this analysis does help point out an important way in which NOA might be thought to be *unsatisfying*.[8] It requires that we say nothing more at a point where it makes us profoundly uncomfortable to do so. If there is indeed something general (and defensible) to be said about the kind of truth scientific claims (or important categories of them) have, or the general aims they can reliably achieve, or something similar, we surely want to be able to say it. Thus perhaps what we should hesitate to embrace is Fine's (1991, 93) suggestion that "the ambiguity over the character of acceptance in science that results from not raising the realism/instrumentalism question seems to be an ambiguity we can quite well live with" while we "get on with other things." And perhaps this encourages us to regard NOA as a counsel of despair or last resort to be embraced with resignation if and when we finally come to agree that there is nothing both general and defensible to be said in this connection and after exhausting (at least) the realist and instrumentalist alternatives. Of course such reluctance and hesitation will seem mysterious to those who follow Fine in finding something "natural" about the attitude NOA expresses. I myself find it natural to wonder whether there is not something general to be said about the kind (or kinds) of truth that independently identifiable categories of important scientific claims have (just as I do about the more homely and familiar claims of everyday life), and I therefore find NOA's robust minimalism disappointing, but perhaps with sufficient philosophical therapy I could get over it. Until then, whenever "NOA whispers the thought that maybe we can actually get along without extra attachments to science at all" (Fine 1986b, 172), I will have to simply whisper back, "But I would prefer not to."

8. Indeed one of Fine's claims is that NOA shows how minimal an adequate philosophy of science can be, and this is where I think Musgrave's argument should really lead him to dissent (as perhaps he recognizes in glossing his accusation of know-nothingism as pointing out just how minimal NOA really is).

This response to Musgrave's analysis helps illuminate another important feature of Fine's position that has invited criticism of NOA from a fellow traveler on the quietist path. As Fine's remark about "the ambiguity...that results from not raising the realism/instrumentalism question" suggests, NOA recognizes a genuine *distinction* between realist and instrumentalist interpretations of science. Moreover, it even seeks to recapture something very like this distinction as a description from within the practice of science itself of the diversity of the various attitudes that the evidence can lead us to take toward particular scientific claims (or that our acceptance of a scientific truth in the "homely" sense can amount to) in particular sets of circumstances: "One NOAer might even find specific grounds in certain cases for bracketing belief in favour of commitment, for instance, while another might go for some measure of belief" (Fine 1986b, 176–177). But Blackburn criticizes NOA on this score—for as we have already seen, he does not think this distinction can be drawn in a coherent way at all. As he says it: "By allowing that there is a difference here, albeit one that is invisible in the practice of science, Fine underestimates the problem. It is not that we acknowledge the difference, but then stop worrying about it. It is that we have lost sight of a difference to acknowledge" (Blackburn 2002, 128).

Instead of turning our backs on the *interest* or *importance* of the distinction between instrumentalist acceptance and realist belief, Blackburn insists that there is no such distinction to turn our backs on in the first place. And this suggests quite a different form of quietist response to the successes individual scientific theories enjoy:

> Suppose my practice is successful: my space rockets land where and when they should. What is the best explanation of this success? I design my rockets on the assumptions that the solar system is heliocentric, *and it is*. Why is our medicine successful? Because we predicted that the viruses would respond in such-and-such a way, *and they do*. In saying these things we...are simply repeating science's own explanation of events. There is no better one—unless there is a better scientific rival.... In other words there is no difference between explaining the gas laws in terms of the kinetic theory of gases, and explaining why we do well relying on the gas

laws, which has to be done in the same terms. There is no getting *behind* the explanation. (Blackburn 2002, 130)

Like Fine, Blackburn thinks that there is nothing more to say about the sense or senses in which the accepted claims of science are true. We do not go on, for instance, to say that what we mean in saying that "the solar system is heliocentric" is true is that this claim "corresponds" to the way the world is, or that the success of the heliocentric system demands the truth of this theory as its only plausible explanation, or any such thing. But unlike Fine, he also thinks that after we have accepted science's own explanation of why a given scientific belief is successful, there is simply nothing more to say, *period*: there is not even any room "behind" such an explanation that we might disdain to occupy by refusing to take a stand on whether we accept it in a realist or an instrumentalist spirit.[9]

As it turns out, then, we cannot avoid answering the question of whether there is any coherent distinction between realism and instrumentalism to be made out even if we hope to decide no more than what *kind* of quietist to be. Quietists as well as instrumentalists and realists will thus need a convincing answer to this question, as do we all if we are to know what (if anything) is at stake in adopting a realist, instrumentalist, or quietist line on the interpretation of scientific theories. Let us return once again, then, to the putative contrast between realism and instrumentalism and ask whether the latter position can be formulated in an independently attractive way that leaves room for an important distinction of the sort that Stein and Blackburn have suggested cannot be coherently drawn.

4. RECLAIMING INSTRUMENTALISM

The fundamental intuition that there is an important difference between believing a scientific theory and merely using it to achieve our practical

---

9. Intriguingly, then, while Fine (1986, 175) embraces what he calls "quietism" and remarks that Blackburn (1984) treats the view "dismissively," Blackburn (2002) ultimately embraces a quietism that is, if anything, even quieter than Fine's own.

goals and objectives is not far to seek. We routinely use Newtonian mechanics to send rockets to the moon, for example, but no one believes that Newtonian mechanics is even approximately true in the sense that matters for the debate over scientific realism. When it comes to the sorts of fundamental claims about the constitution of nature that the scientific realist thinks must be at least probably and/or approximately true in the case of our best contemporary theories, Newtonian mechanics is acknowledged on all sides to be radically false and yet nonetheless an extremely powerful and reliable tool for predicting the behavior of objects like billiard balls, cannonballs, and planets under a wide (though not unrestricted) range of conditions.

Let us begin, then, by asking what the *realist* means when *she* claims that Newtonian mechanics is a radically false theory that nonetheless serves as a useful instrument. On the one hand, she simply rejects the Newtonian account of the constitution of nature: she denies that gravity is a force exerted by massive bodies on one another, that space and time are absolute, and so on. But on the other hand, she knows perfectly well how a Newtonian would apply the theory to make predictions about and intervene with respect to entities, events, and phenomena *as they are given to her by other theories that she does believe*. That is, while the relativity theorist has an independent account of the cannonballs, inclined planes, and rockets to which she could apply the theoretical mechanics she believes to be literally true, she also knows how a Newtonian would apply her own distinctive theory to characterize those same cannonballs, inclined planes, and rockets in terms of the masses, forces, collisions, and so forth that would allow her to predict and intervene with respect to them, so described. This has nothing to do with the fact that cannonballs, inclined planes, and rockets are observable entities: she knows equally well how a Newtonian would identify forces and masses so as to make predictions about the gravitational motions of subatomic particles. And over whatever domain she thinks the theory is (more or less) instrumentally reliable she can make use of it, because she knows how to apply it like a Newtonian would to entities, events, and phenomena whose existence she countenances independently of the theory.

It may come as a surprise to note that this same fundamental strategy is also available to the instrumentalist. That is, a scientific theory demonstrates its instrumental utility in application to features or aspects of the world that we take ourselves to have some grasp of or route of epistemic access to that is independent of the theory itself. Thus in place of the instrumentalist's usual attempt to separate a theory's claims into those we must believe to make effective instrumental use of it (e.g., its claims about observables or its predictions and recipes for intervention) and those we need not, we might instead allow that an instrumentalist about a given theory simply believes *all* the claims of that theory *as those very claims can be understood independently of any theory or theories toward which she adopts an instrumentalist stance.* Take, for example, the claim that "a liquid supercooled below its standard freezing point will crystallize in the presence of a seed crystal or nucleus." This is a highly theory-laden claim, but it grounds a specific set of implications and expectations concerning entities, events, and phenomena *as we understand them independently of contemporary chemical theory.* We know how to identify freezing points, create supercooled liquids and seed crystals, and recognize crystallization using procedures that could be followed by someone without any substantial knowledge of contemporary chemical theory—including the procedures we would use to exhibit these entities and processes to a neophyte in the laboratory (typically while providing a complementary explanation of what contemporary theoretical orthodoxy holds is going on).

This proposal for regaining instrumentalism is broadly sympathetic in spirit with the suggestion (found, for example, in Kitcher 1978, 1993; Stanford and Kitcher 2000) that the reference of natural kind terms in scientific contexts is fixed quite differently on different occasions of use in ways that are sensitive to the dominant referential intention of the term's user on that occasion.[10] On some occasions, that is, Joseph Priestly's tokens of "dephlogisticated air" referred to the substance whose inhalation had made his breathing particularly light and easy afterward or the substance he "exploded together" with "inflammable air" to produce water or nitric acid, while the references of some other tokens of this very same term were instead fixed on the relevant

---

10. For more details regarding this connection, see Kitcher 1993, esp. chap. 4; and Stanford 2006, chap. 8.

occasion of use by Priestly's dominant intentions to satisfy the theoretical description "air from which the substance emitted in combustion has been removed" and thus failed to refer at all. Here we simply note further that the dominant referential intention that allowed Priestly's genuinely referring tokens to do so were those that picked out the referent of "dephlogisticated air" in a way that relied on some further route of independent epistemic access, through his senses or through further beliefs about dephlogisticated air that did not depend in any fundamental way on the commitments of phlogistic chemistry (e.g., "the stuff in this jar," "the stuff that makes my breathing so light and easy").

Although this account of the matter has no use for the idea of a pure observation language or a foundational epistemic role for observability as such, it recognizes that our various sensory modalities will be among the routes most commonly used by us to secure an independent epistemic grasp of entities, events, and phenomena described by the theories toward which we adopt instrumentalist attitudes. For such an instrumentalist, a scientist who finds a new way of detecting an entity or phenomenon or its causal influence or of creating it in the lab gives us a new route of epistemic access to that entity or phenomenon that need not and often will not depend on the commitments of the theory that sent us looking for it in the first place.

It is not, however, a grasp or understanding of an entity or phenomenon that relies on the senses alone that instrumentalism of this sort requires; instead it simply requires that we recognize sources of knowledge about entities, events, and phenomena in the world that are independent of any particular theories toward which we adopt an instrumentalist attitude. Moreover, this remains a perfectly coherent possibility even if we ultimately decide that *all* our knowledge of the world is theoretical in character. If we suppose that W. V. O. Quine (1976) was right to suggest that the familiar middle-sized objects of our everyday experience are no less "theoretical" entities hypothesized to make sense of the ongoing stream of experience than are atoms and genes and that it is by means of theories in this broadest sense that we come to have *any* picture at all of what our world is like, we will then simply be faced with the need to decide precisely *which* theories we will regard merely as useful conceptual instruments (as well as why all and only those particular theories). To be sure, on this account instrumentalism is not an attitude that can be coherently adopted toward all theories whatsoever (at least not if Quine is right), but the example of Newtonian mechanics already illustrates why even scientific realists will need to decide which theories they are instrumentalists about on the basis of individualized consideration of those theories instead. Even the scientific instrumentalist will adopt a realist attitude toward a wide variety of hypotheses concerning what Quine called (e.g., 1976, 250) "the bodies of common sense", and the epistemic features or deficiencies that determine how broadly she will apply the instrumentalist attitude in other cases will surely depend on what her reasons are for adopting that attitude toward any theories in the first place.[11]

Of course if we are instrumentalists about a theory that makes claims about entities or events to which we have no routes of independent epistemic access, our belief in those claims "as they can be understood independently of the theory" will be empty. Because contemporary particle physics does not allow quarks to be isolated, for example, it posits the existence of gluons to bind quarks within a proton, but we have no routes of independent epistemic access to gluons beyond positing them to fulfill this hypothesized function. Those who are instrumentalists about particle physics, then, will not believe *any* of its claims regarding gluons but will simply make use of them in the process of generating further claims whose content is specified by sources of knowledge (including other theories) toward which she does not adopt an instrumentalist attitude.

Indeed on this account the fundamental difference between the realist and the instrumentalist turns out *not* to be a matter of their respective epistemic attitudes toward "theories" or theoretical knowledge as such.

---

11. It is this form of instrumentalism for which I tried to provide a principled foundation in Stanford 2006, whose final chapter contains a more detailed development of the sort of instrumentalist position I have briefly sketched here. Throughout that work I argue that the crucial feature of any theory that should provoke us to adopt an instrumentalist attitude toward it is its demonstrated vulnerability to the probable existence of unconceived alternatives that are also well confirmed by the evidence. I also argued (perhaps a bit too quickly; see Stanford 2011) that this includes most if not all of our fundamental scientific theories.

Both realists and instrumentalists regard some theories merely as useful instruments for predicting and intervening with respect to a given natural domain but not probably or approximately true descriptions of the domain in question, and those traditionally described as instrumentalists are simply willing to take this view of a much wider range of theories than their realist counterparts. The remaining difference would seem to consist simply in the fact that the instrumentalist is sometimes willing to regard even an extremely powerful and pragmatically successful theory as no more than a useful instrument *even when she knows of no competing theory that she thinks does indeed represent the truth about the relevant natural domain*. But there seems little reason to think that this difference turns what looks for all the world like a clear and coherent distinction between the scientific realist's own attitude toward Newtonian mechanics and her attitude toward the special and general theories of relativity into a muddle.

To be sure, it is a difference that makes a difference. When pressed, the realist will use either special or general relativity and not Newtonian mechanics to make predictions when she thinks minute differences might be important. She will use relativity and not Newtonian mechanics to ground her further inquiry into and exploration of the domain of nature described by both theories. And she will use relativity and not Newtonian mechanics to extend her theoretical representation of nature, her predictions, and her interventions into new parts of the relevant theoretical domain where the approximate predictive equivalence of the two theories is either unknown or is known to fail. The instrumentalist can do none of this. Instead, she must admit that she has no choice but to use a theoretical instrument she does not believe to be true when she makes predictions, no matter how high the cost of even minute errors. She must also admit that she has no better point of reference from which to initiate further inquiry into new aspects of a given domain of nature than the most powerful instrument she knows of for thinking about it and/or related natural domains. And she must admit that her only route forward in extending our powers of pragmatic engagement in a given natural domain is to press on in trying to apply the best existing conceptual tool she has to that domain to see whether and/or how it succeeds or fails, finding only by painful trial and error the situations and circumstances that suggest where existing tools require improvement and even what the radically different alternatives that would perform even better might be.

The form I have here given instrumentalism is importantly connected to the second of the distinctions Blackburn cites as a source of ersatz intuitive plausibility for a difference between instrumentalism and realism: the difference between *fully* embracing (or fully "immersing" oneself in) a theory and doing so in a more cautious or tentative spirit. Unlike the realist (and even van Fraassen's constructive empiricist), the instrumentalist described here believes that even our best current conceptual tool for thinking about a given scientific domain will *ultimately* be found wanting and be replaced by an alternative that offers even more powerful resources for guiding our pragmatic engagement with nature. Thus she has a very different picture of the future of scientific inquiry than the realist does, and she will be commensurately less willing to reach a conclusion that depends on the long-term stability of our best scientific theories than will her realist counterpart.

Furthermore, although she knows of no better way to extend her investigation into a novel domain of nature or an unfamiliar context than to explore and test the predictions of her best scientific theory or theories about it, we should expect the instrumentalist to be systematically less willing than her realist counterpart to *trust* those predictions until the theory's predictive adequacy in that unfamiliar domain or novel context can be independently verified. This does not mean, however, that she will view the theory's predictions as no more likely to be true than predictions somehow generated at random. After all, her best scientific theory enjoys a successful track record in guiding our pragmatic engagement in applications and/or domains of nature continuous with or closely related to one in which she is trying to extend it (cf. Stanford 2000).[12] Of course a sensible caution about novel applications of our scientific theories can be motivated for the realist by worrying about the complexities and/or unknown interactions involved in applying a given

---

12. Of course in the absence of any such successful track record, she might be even more circumspect about a theory's predictive prospects, but then so might (and should) her realist counterpart.

theory in a new domain or context, but there is no reason to think that the degree or character of such caution will or should precisely mimic that motivated by the instrumentalist's much more general lack of trust in the truth of the theory's fundamental claims or any reason to think that the circumstances in which these two sorts of caution respectively become acute will ultimately be the same.

Perhaps most importantly of all, however, we should expect the instrumentalist to be much more open than her realist counterpart to the serious exploration of fundamentally distinct alternatives to her best current scientific theories (though not necessarily any particular alternative of course). For the realist, the exploration of such alternatives is rendered reasonable only by her judgment that it is only *probable* that her best current theory is true, and such exploration is a sensible investment of time and effort only in the small space (and to the limited extent) left open by the unlikely prospect that it is false. By contrast, the instrumentalist doubts that even the best presently available conceptual tools she has for thinking about nature will retain that status indefinitely as future inquiry proceeds, and this will lead her to be systematically open to the thoroughgoing exploration of fundamentally distinct alternatives, at least one of which she fully expects to replace her best current theory in the fullness of time.[13] Thus my instrumentalist does not say,

---

13. Of course this represents a judgment about the fruitful or legitimate directions and investment of efforts for scientific inquiry as a whole, not a prediction about how any individual scientist will or should spend her time, which presumably owes as much to personal taste and expertise as anything. Nonetheless, in this important sense the instrumentalist is much more fully committed than the realist to the open-ended character of scientific inquiry itself. I do not mean to suggest, however, that one or the other view more faithfully represents the attitudes of scientists themselves on this question: it seems to me that scientists hold widely varying views on this matter. More importantly, even if we became satisfied that one or the other attitude toward further inquiry really were more widely represented among practicing scientists, this would not automatically be an important consideration in favor of adopting it and/or regarding it as a better interpretation of the institution of science *itself*. In this connection scientists simply offer candidate interpretations of the institution in which they participate, on all fours with other contenders, and they speak with extensive knowledge of the institution but no special authority about its proper interpretation, just as (it has been suggested; cf. Dworkin 1986) a knowledge of what an individual legislator hoped to accomplish by voting for a particular law need not carry any special authority in determining how the

with Blackburn's, "I don't really believe in genes, or atoms, or gluons; it is just that I structure all my thoughts by accepting that they exist." Both her thoughts and her further inquiry into nature are structured quite differently than they would be if she believed that our best current theories of inheritance or of the minute constitution of matter were true. And notice also that the instrumentalism articulated here dissents not only from realism but also from the "core position" that quietism embraces and suggests that realism and instrumentalism hold in common: in the various ways described above, our instrumental commitment to the claims of theoretical science is simply *not* like our embrace of the more homely and familiar truths of everyday life.

Will the instrumentalist be unjustified or irrational in thus making use of a theory she does not think is even probably or approximately true and that she ultimately expects to fail in application to new situations and in new contexts? I do not see why. This is irrational or unjustifiable only if she knows of some particular alternative that she expects to give better results in a given concrete application or to provide a better guide to her further inquiry, and of course this is just what she does not have. Moreover, though she expects that even the best conceptual or theoretical instrument presently in her possession will ultimately fail, she has no specific reason to think that the very next occasion she has to apply it will be the one on which its fundamental inadequacy will be revealed. In this she occupies a position something like that of a person who is trapped underground and whose only route of escape is to make use of a blasting device with an unknown but limited number of charges: if she really has no options, she must use the device she knows will ultimately fail and simply hope on each crucial occasion that it is not the one on which that failure will materialize. Each is *serially* rational in using an instrument or tool that she knows to have a fair chance of working on that particular occasion, and each is faced only with the alternative of doing nothing at all. All that either can do *because* she knows that the best tool she has will

law itself ought to be interpreted or applied to future circumstances. And in this difference between achieving a better fit to the attitudes of its practitioners and a better fit to the practice and/or history of the institution itself lives the most important room for the philosophy of science to play a legitimate normative role in guiding scientific inquiry.

ultimately fail is to be cautious about assuming it has worked on a new occasion (i.e., in a new context) until this can be independently verified and to keep a weather eye out for other, potentially even better tools to use. In all of this perhaps a person trapped underground and using tools of unknown reliability to try to escape bears an uncomfortable but uncanny resemblance to the rest of us.

References

Blackburn, Simon. 1984. *Spreading the Word*. Oxford: Clarendon.
———. 2002. "Realism: Deconstructing the Debate." *Ratio* 15:111–133.
Dworkin, Ronald. 1986. *Law's Empire*. Cambridge, MA: Belknap.
Fine, Arthur. 1986a. *The Shaky Game: Einstein, Realism, and the Quantum Theory*. Chicago: University of Chicago Press.
———. 1986b. "Unnatural Attitudes: Realist and Instrumentalist Attachments to Science." *Mind*, n.s., 95:149–179.
———. 1991. "Piecemeal Realism." *Philosophical Studies* 61:79–96.
Horwich, Paul, 1991. "On the Nature and Norms of Theoretical Commitment." *Philosophy of Science* 58:1–14.
Kitcher, Philip. 1978. "Theories, Theorists, and Theoretical Change." *Philosophical Review* 87:519–547.
———. 1993. *The Advancement of Science: Science without Legend, Objectivity without Illusions*. New York: Oxford University Press.
———. 2001. "Real Realism: The Galilean Strategy." *Philosophical Review* 110:151–197.
Laudan, Larry. 1981. *Science and Hypothesis: Historical Essays on Scientific Methodology*. Boston: Reidel.
Musgrave, Alan. 1989. "NOA's Ark—Fine for Realism." *Philosophical Quarterly* 39:383–398.
Nagel, Ernst. 1961. *The Structure of Science*. New York: Harcourt, Brace, and World.
Quine, W. V. O. 1976. "Posits and Reality." In *The Ways of Paradox and Other Essays*. 2nd ed. Cambridge, MA: Harvard University Press.
Sklar, Larry. 2000. *Theory and Truth*. New York: Oxford University Press.
Stanford, P. Kyle. 2000. "An Antirealist Explanation of the Success of Science." *Philosophy of Science* 67:266–284.
———. 2006. *Exceeding Our Grasp: Science, History, and the Problem of Unconceived Alternatives*. New York: Oxford University Press.
———. 2011. "Damn the Consequences: Projective Evidence and the Heterogeneity of Scientific Confirmation." *Philosophy of Science* 78:887–899.

Stanford, P. Kyle, and Philip Kitcher. 2000. "Refining the Causal Theory of Reference for Natural Kind Terms." *Philosophical Studies* 97:99–129.

Stein, Howard. 1989. "Yes, but…Some Skeptical Remarks on Realism and Anti-realism." *Dialectica* 43:47–65.

van Fraassen, Bas. 1980. *The Scientific Image*. Oxford: Clarendon.

Chapter 5

Structure and Logic

SIMON SAUNDERS AND KERRY MCKENZIE

> If any problem in the philosophy of science justifiably can be claimed the most central or important, it is that of the nature and structure of scientific theories.
>
> Frederick Suppe, *The Structure of Scientific Theories*

1. INTRODUCTION

Structuralism of one sort or another was a dominant force in philosophy of science throughout much of the last century and along with it logic—by which we mean formal methods, both in syntax and semantics (or model theory). Logic in this general sense might be thought to include set theory, and structuralism itself was often made out in terms of set theory. This chapter is about structuralism as implemented through theory reformulation in formal logic and set theory.

Structuralism thus construed was central to logical empiricism: did it not end with it? Oddly, no: the quotation by Frederick Suppe is from *The Structure of Scientific Theories* (1977), which was devoted, by and large, to the demise of logical empiricism (to the end of the "received view" in philosophy of science). For structuralism in our sense could be variously motivated—even by considerations that had nothing to do with rational reconstruction in Rudolf Carnap's sense and even in contexts that were indifferent or even opposed to formal logic as the language of science. Thus post-Kuhn, a structuralist approach to theories still seemed the right response in the face of the repeated history of theory change, reminiscent of the view of Henri Poincaré, and it seemed the right way to cope with the metaphysical vagaries of quantum mechanics – where by most accounts the traditional concept of an object is in trouble. With the move away from the syntactic approach to theory formulations, set theory emerged as the natural candidate to define theories in terms of collections of models. And not all were agreed that syntactical method was a mistake in philosophy of science: "structural realism," at least in the form John Worrall gave it, used the same formal methods as did Carnap. Only the appeal to conventionalism was changed.

But these renewals are a mistake: structuralism, in the sense we are construing it, is at a dead end. There are, to be sure, broad structuralist lessons to be drawn from physics, but that is another story: it has little to do with structuralism as theory reformulation in terms of logic and set theory.

The greater part of what follows is about this history. It needs retelling, for the motives for structuralism are diverse and only some of them are still worth having. Few of them make sense independent of the project of theory reformulation. It is the latter that is most entrenched in philosophy of science—Suppe's point is well taken—but it should be abandoned all the same. Whether what remains of the structuralist project is worthy of the name is at this point an open question.

2. NEO-KANTIANISM

Logical empiricism began with neo-Kantianism and by some lights hardly went beyond it. Neo-Kantianism at the turn of the last century was embraced by physicists and philosophers alike. It assumed without question the objectivity and authenticity of scientific knowledge and aimed to account for how mental representations might conform to extramental objects consistently with the subjectivist picture of experience and judgment. It was the concept of structure that was thought to supply the answer.

Hermann von Helmholtz, Heinrich Hertz, Max Planck, and Poincaré were its most prominent exponents among physicists, Bertrand Russell and Ludwig Wittgenstein among philosophers. They were agreed that the content of any claim about the world that transcended a description of structure must rest either on indexical elements or on acquaintance with perceptual qualia, neither of which can provide the intersubjective, communicable meaning taken to be characteristic of science. As such it was opposed to positivism in Ernst Mach's sense, in which the further removed from sense and sensibility, the further science departed from truth. While Planck spoke of the need to remove the element of human sensibility from our description of the world altogether, Helmholtz had earlier argued that there is a formal element of experience that may serve as a reliable guide to the noumenal in the neo-Kantian sense. He and later Hertz supposed that the perceptual channels through which knowledge was mediated yielded pictures of the world—*Bildtheorie*—whose structure corresponded to that of the world (the "external" world). Structural similarity, as one-to-one correspondence, was the neo-Kantian alternative to the immediacy of Kantian intuition, and structure as made out in Fregean, logical terms was the alternative to the schematism of intuition by the categories of Immanuel Kant's transcendental logic. As Russell (1927, 254) put it in *The Analysis of Matter*, to the question of what our knowledge of the external world consists in, "the answer is structure; and structure is what can be expressed by mathematical logic."

Formal method was thus part and parcel of the idea of structuralism, but it quickly opened the door to the charge of triviality. The objection was due to Maxwell Newman: let the structure of the external world be W and suppose that this structure alone is known; then, in Newman's (1928, 144) words, "any collection of things can be organized so as to have the structure W, provided there are the right number of them." That is, given that there is the right number of objects, any set-theoretic statement about the world can be made true simply by viewing the world as composed of the right collections of sets in the arrangement of W (we shall shortly restate this argument in more detail). It seems therefore that if structuralism is true, only cardinality questions are open to empirical checks: anything further can be made true a priori. Russell granted the point and never returned to defend his original claims regarding perceptual knowledge.

As a general epistemic thesis in the form Russell gave it, structuralism was clearly unsustainable. But other varieties, less wedded to neo-Kantianism and its dichotomy of internal and external worlds, were thought to bypass Newman's objection. As we shall see, however, that hope was forlorn; if anything, the difficulty only became more general.

3. CARNAP'S STRUCTURALISM

A high point of logical empiricism and one of Carnap's most influential papers was "The Methodological Character of Theoretical Concepts" (1956). There Carnap divides theories into languages and theory formulations. Languages are further divided into sublanguages: an observation language V_O, which consists of observation terms and logical vocabulary, and a theoretical language V_T, which consists of theoretical terms and (a richer) logical vocabulary. With regard to theory formulation, the laws of a theory are supposed to be expressed as a list of theoretical postulates T, which is assumed to be finite and which, given the assumed richness of the formal language, is assumed available no matter how mathematical the theory in its textbook formulations. To T are joined correspondence rules C that, again, are assumed to be finite and that relate the theoretical terms to the observation terms. These predicates are such that any given place of the predicate is always taken by a term from either the class V_O or the class V_T. The theory itself is conceived of as the conjunction TC.

Despite its centrality to their position, the logical empiricists never actually wrote down such a theory. Instead, they took it for granted that theories could be consistently, realistically, and practically formulated in this highly formalized way. It was reasonable, perhaps, to postpone the hard work of rational reconstruction until its purpose was more fully established.

Questions of meaning were at the heart of this purpose. In particular the positivists' concern lay with "the problem of meaning" of theoretical terms, considered to be two-pronged. First, there was the issue that such terms are phenomenologically removed, "not meaningful in the same way that observation terms...are meaningful", and thus judged to be of unsettled cognitive status (Carnap 1966, 248). What was needed was a criterion of how such terms could be meaningful in spite of this, and preferably one that did not simultaneously sanction the meaningfulness of purely metaphysical statements. What did seem clear was that any theoretical term "derives its meaning from the context of a theory", and also that what distinguished such terms in scientific theories from their metaphysical counterparts was the fact that they were "connected to laboratory observables" (Carnap 1966, 238). As such their meanings are holistic and theory dependent—as seems entirely appropriate from a post-Kuhn, post-Quine perspective—but their contact with experience implies a crucial connection with correspondence rules. However, theoretical terms could not be defined by such rules, because they must always "be open to modifications by new observations" (Carnap 1966, 239) by new laboratory devices. It was capturing this open-endedness of correspondence rules that constituted the second problem connected with meaning. Given this feature, theoretical terms are only "partially interpreted" by the theory, even though they derive their meaning from theory. What, then, is the missing meaning of theoretical terms, which is continually gained upon?

Instead of answering this "puzzling question," Carnap (1966, 249) noted that there was a way it could be side-stepped: one can switch to a formulation of the theory that simply eliminates the very terms about which these questions are raised, while preserving many essential features of the theory. The formulation that achieves this was called (in Hempel 1958) the *Ramsey sentence* R^{TC} of the theory TC.

The Ramsey sentence is formed by existentially generalizing away every theoretical term in TC; that is, denoting such terms by $T_k \in V_T$, $k = 1, \ldots, N$ and observation terms by $O_j \in V_O$, $j = 1, \ldots, M$ and writing TC as

$$TC = TC(O_1, \ldots, O_M; T_1, \ldots, T_N).$$

The Ramsey sentence of the theory is

$$R^{TC} = \exists \psi_1 \ldots \exists \psi_N TC(O_1, \ldots, O_M; \psi_1, \ldots, \psi_N)$$

where each Ψ_k is a variable of the appropriate type and arity. Assuming that the original theory TC is a first-order theory, R^{TC} is evidently a second-order construction. It qualifies as a structuralist expression of theoretical knowledge, because the entire content of the theoretical system is reduced to set-theoretic structure—not knowledge of specific properties and relations in the world but of any properties, any relations, that interconnect with one another in the right way. No theoretical meaning apart from this structure remains: the description of this theoretical system is entirely in logical terms.

This was the position on which Carnap eventually settled, and he linked it to his final response, worked out in the late 1950s, to Willard Van Orman Quine's attacks on the analytic-synthetic distinction. In this way he expanded on what could positively be said about partially interpreted terms: theoretical terms contribute to theories only through their logical type, their arity, and their place in the structure of a theory. Thus theoretical relations were captured but only up to isomorphism. Carnap (1966, 254) supposed that his earlier mistake was to use interpreted terms as opposed to predicate variables, which "presupposes that theoretical terms, such as 'electron' and 'mass', point to something that is somehow more than what is supplied by the context of the theory. Some writers have called this the 'surplus meaning' of a term." Such surplus meanings are now nowhere to be seen. Each theoretical term has been replaced with a quantified second-order variable, which "does not refer to any particular class. The assertion is only that there is at least one class that satisfies certain conditions" (Carnap 1966, 251). But R^{TC} was by no means judged to lack cognitive content as a result of such indefinite reference. On the contrary, the Ramsey sentence was judged to express the entire factual content of the original theory (Carnap 1966, 270). What underwrites this interpretation is the fact that "any statement about the real world that does not contain theoretical terms—that is, any statement capable of empirical confirmation—that follows from the theory will also follow from the Ramsey sentence. In other words, the Ramsey sentence has precisely the same *explanatory and predictive power* as the original system of postulates" (Carnap 1966, 252).[1]

---

1. The proof that the Ramsey sentence for a theory has the same empirical consequences as the theory itself is given in Ramsey 1931, 212–215, 231.

The connection with the analytic-synthetic distinction was this: since the Ramsey sentence expressed the factual content of the theory but TC contained the theoretical terms, it was plausible to declare the sentence $R^{TC} \to TC$ the "meaning postulate" (also called the Carnap sentence) for the language of the theory TC. It is the weakest sentence that can be added to R^{TC} from which the original theory can be recovered. Thus R^{TC} expresses the empirical content of TC, while the Carnap sentence $R^{TC} \to TC$ expresses its meaning.

4. REALISM AND INSTRUMENTALISM

Given such an emphasis on the role of the transformed theoretical terms in deriving observation statements, the Ramsey sentence was sometimes assumed to be of wholly instrumentalist import. That in any case was the view targeted by Carl G. Hempel in "The Theoretician's Dilemma" (1958). Newman's objection went unmentioned, but Hempel's conclusion, that the Ramsey sentence R^{TC} at least involves the existence claim that there are sets and relations among individuals over and above the individuals themselves, is a response to it of sorts.

Carnap's own view, as with existence claims more generally, was equivocal. But he seemed at least to agree that the choice of the Ramsey sentence as the definitive expression of theories did not amount to an expression of anything in conflict with realism:

> In Ramsey's way of talking about the external world, a term such as "electron" vanishes. This does not in any way imply that electrons vanish, or, more precisely, that whatever it is in the external world that is symbolized by the term "electron" vanishes. The Ramsey sentence continues to assert, through its existential quantifiers, that there is something in the external world that has all those properties that physicists assign to the electron. It does not question the existence—the "reality"—of this something. It merely proposes a different way of talking about that something. The troublesome question it avoids is not, "Do electrons exist?" but, "What is the exact nature of this electron?" (Carnap 1966, 252)

How to characterize the domain of "these somethings," which are affirmed to exist and serve as the range of the Ramsey variables, was left carefully ambiguous. But what matters for our purposes is that, whether abstract or concrete, R^{TC} is still supposed to express *factual*, that is, synthetic or a posteriori, truth. Thus even when the variables range over abstract objects, such as numbers, according to Carnap (1963, 963), the Ramsey sentence says that "the observable events in the world are such that there are numbers, classes of such, etc., which are correlated with the events in a prescribed way and which have amongst themselves certain relations; and this assertion is clearly a factual assertion about the world."

Whatever range is chosen, as an expression in second-order logic the semantics of the Ramsey sentence requires the use of second-order interpretative structures. These structures contain two types of domain: a first-order domain, as in first-order logic, and a second-order domain composed of sets of elements of the first-order domain. There are furthermore two types of structure: *full* structures, which consist of the full power set of the first-order domain, and *Henkin* structures, in which the second-order domain is a proper subset of the full structure. It is full structures alone that are thought to pertain to "real" second-order logic: hence the alternative designation of such structures as "standard." Given that Carnap intended the language of the theory to be sufficiently rich to express any relation whatsoever, it would seem that the domain interpreting his Ramsey structure should be full. But this has an immediate consequence for the status of the theoretical claims of R^{TC}. Despite Carnap's claim that there is some "correlation" that results in the "factuality" of the theoretical system, it follows as with Russell's system that the nonobservational propositions of R^{TC} reduce without residue to a claim about the cardinality of the first-order domain: any further characteristics of theoretical relations, both intrinsic and extrinsic, follow a priori.

It is easy to see why the Ramsey sentence reduces theoretical claims to triviality.[2] We will assume that the extensions of the theoretical and

---

2. The following is adapted from Melia and Saatsi 2006.

observation terms are completely disjoint (though we will revisit this assumption below); we will also assume for simplicity that there are no "mixed" predicates (see Ketland 2004 for the more general presentation). Start with an un-Ramseyfied theory *TC*, which we take to contain as its nonlogical terms the observation predicates $O_1, \ldots, O_M$ and theoretical predicates $T_1, \ldots, T_N$. Thus $TC = TC(O_1, \ldots, O_M; T_1, \ldots, T_N)$, and assume that *TC* is both consistent and empirically adequate. It follows from consistency that *TC* has an "abstract" model, and let this model be $M = ((D_1, D_2); O^1, \ldots, O^M; T^1, \ldots, T^N)$, where D_1 is the first-order domain for the observational vocabulary, D_2 the first-order domain for the theoretical vocabulary, the $O^1, \ldots, O^M$ are the interpretations of the predicates $O_1, \ldots, O_M$ and the $T^1, \ldots, T^N$ the interpretations of $T_1, \ldots, T_N$.[3] Since *TC* is by assumption empirically adequate, the domain D_1 just is the set of observable entities, which we may as well denote D_O, and each set of ordered tuples O^j in the model has as its members all and only those tuples of observable objects that stand in the relation expressed by O_j.

Now consider the "intended" domain of the theoretical vocabulary—the set of theoretical entities—and let this set be D_T. So long as D_2 is of the same cardinality as D_T, we can define a 1:1 mapping φ that maps elements of D_2 to D_T while mapping each member of D_O to itself. Then $M^t = ((D_O, D_T); O^1, \ldots, O^M; \varphi(T^1), \ldots, \varphi(T^N))$ is isomorphic to M, from which it follows that M^t is a model for *TC*. By the clauses for existential generalization in standard second-order model theory, $\exists \Psi_1 \ldots \exists \Psi_N T(O_1, \ldots, O_M; \Psi_1, \ldots, \Psi_N)$ is also true in (D_O, D_T); but this is the theory's Ramsey sentence. Thus all we need demand of the intended interpretation of a Ramsey sentence, beyond empirical adequacy, is that the first-order domain contains sufficiently many elements as are needed to make it true in an abstract model. It follows that its a posteriori content concerning unobservables is reduced to a claim about cardinality. This is the Newman objection as applied to Ramsey sentences.

It will be clear that if the second-order domain formed of sets of members of D_T does not contain every relation that can be defined on D_T

---

3. In keeping with standard second-order model theory, the second-order domain is not included explicitly, as it is understood that it consists of the full power set of the first-order domain and thus determined by it.

then the argument will not go through, for the argument makes essential use of the assumption that the sets mapped to under the arbitrary one-one mapping φ are available as values of variables. In the case where the second order domain is not full, one would thus require additional information to ascertain that the sets mapped to by φ are indeed in it. But since the whole solution to the problem of articulating the notion of partial interpretation lies in the claim that theoretical relations are captured only up to their logical type and arity, it seems that there is in principle no way to isolate and exclude any relation in its isomorphism class except on arbitrary grounds. That such a privileging of particular relations in an isomorphism class is incompatible with structuralism was also pointed out by Newman. (We will return to this issue below.)

It thus appears that the trivialization of claims about unobservables is an inevitable consequence of the assumption that our knowledge of theoretical terms extends only to the level of their structure, as expressed by the Ramsey sentence for the theory in which they occur. The lesson is that if our theoretical knowledge is to be significant in the sense of being sufficiently a posteriori, then our knowledge must consist in more—something that requires more determinate reference than to just any properties and relations with the given structure. But now this "something more"—whatever must be added to R^{TC} to obtain TC itself—makes the difference between a priori knowledge (modulo a cardinality claim) and a posteriori knowledge concerning unobservables. That in itself makes nonsense of Carnap's claim that the "meaning postulate" $R^{TC} \to TC$ is devoid of factual content.

A lesson one may be tempted to draw at this point is that the move to second-order quantification was a bad one: that only a first-order semantics can do justice to realism about theoretical terms. However, it is not so clear that merely passing over to first-order theory formulations is sufficient to make the problem go away. There are of course well-known limitations to the expressive adequacy of any first-order language, but its strengths too in the field of semantics came to be challenged. From these challenges, the idea that knowledge extended no further than structure emerged not as a posit but as a consequence of very general features of correspondence-based referential semantics and, from a realist perspective, a disastrous one.

5. STRUCTURALISM ENFORCED

In "Models and Reality" (1980) Hilary Putnam argued that a triviality similar to that flagged up in the Newman objection afflicted first-order physical theories. His argument was as follows. Consider a theory T to be a formalization of scientific theory: we may consider it to be a formalization of even "ideal" scientific theory—that is, theory in the limit of inquiry that includes every theoretical and operational constraint we care to impose. Let a certain set of sentences be derivable as theorems of T, expressing truths about some set S of observable things and events (here we should err on the side of liberty). Let O-terms denote those predicates that apply to the things and events in S—we now allow that they may apply to unobservables also, but they must apply to at least one member of S (with this we avoid the need for correspondence rules). Let σ_O be an interpretation of T under which the theory is true, so that (among other things) the values in S assigned to each n-ary O-term satisfy all the operational constraints included in T. Thus T under σ_O is empirically adequate, and the interpretation is "intended." However, merely from consistency, T has an abstract model M; since all those sentences required to be true under σ_O are theorems of T, they will come out as true in M as well. Thus M must have a member corresponding to each member of S. But now those members of M of which they are true can all be replaced by the corresponding member of S, and by modifying the interpretation of predicate letters accordingly, one can construct a new model, M^t, in which each term denoting a member of S in the intended interpretation defined by σ_O does denote that member of S. Such a model is therefore *also* an intended model, because it satisfies all the observational and theoretical criteria associated with the intended model. In what sense, then, can T fail to be true, given that it is consistent and true of observable things, when its truth in an intended model follows from its truth about S?

Putnam's conclusion was that the realist position, according to which a scientific theory may yet be false even though it correctly describes every observable thing and event, is ultimately incoherent. For insofar as truth outstrips what is obtained by an understanding and deployment of verificationist procedures in determining truth values for sentences about observable things and events, then, modulo a cardinality constraint, truth comes for free. There is a cardinality constraint to satisfy, here as with Newman's objection, most evidently in the case of finite theories, but—over and above truths about observable things—nothing else is required. For what, over and above an understanding of how reference is made at the observable level, can constrain a model as "intended"? Anything more than this is at bottom simply *unintelligible*, according to Putnam—a correspondence between words and things that is inexpressible, even in the limit of inquiry.

To drive home the point, Putnam compared the situation in philosophy of science to philosophy of mathematics. There too it might be thought that a moderate form of realism can countenance the existence of abstract mathematical objects. But deprive the moderate realist of any epistemic access to this realm, other than proof procedures, and insist that in the limit of inquiry one arrives at a body of mathematical truth that will yet be only a small fragment of this realm; then reference to this realm appears to need some kind of direct intuition of Platonic objects, an "extreme" form of realism in the philosophy of mathematics. Likewise only the extreme, metaphysical form of realism in philosophy of science has the resources to deal with Putnam's objection if truth is not to come cheap. As Putnam (1980, 13) concluded, moderate realism "runs into precisely the same difficulties that we saw Platonism run into."

The point is a serious one, but note that it hardly has the force of a paradox or a skeptical challenge to everyday knowledge. The problem as so far formulated concerns reference to unobservables, as with the Newman objection to the use of the Ramsey sentence in formulating a scientific theory. But in one place in "Models and Reality" Putnam introduced a separate argument based on permuting the extensions of predicates—even observation predicates—that had more drastic consequences.[4] The things to which those predicates applied were permuted in such a way that "corresponding adjustments in the extensions

---

4. In *Reason, Truth, and History* (Putnam 1981) this latter argument moved to center stage. Likewise, Putnam's summary of the conundrums of the metaphysical realist in the preface of his *Realism and Reason* (1985) was formulated solely in these terms.

of all the other predicates make it end up so that the operational and theoretical constraints of total science or total belief are all 'preserved'" (Putnam 1980, 24).

To see that this argument applies to observable things as well, consider the simplified case of a theory T formulated in a language L whose only extralogical symbols are observable predicates (O-predicates) $O_1,..., O_M$. Suppose as before that T is true under an intended interpretation σ with universe of discourse S that assigns to each n-ary predicate O_k the n-ary relation $R_k = (O_k)^\sigma$ on S. Now choose any 1:1 map φ on S such that for at least one relation R_k, $\varphi(R_k) \neq R_k$, where $(u_1,...u_n) \in \varphi(R_k)$ just in case $(\varphi^{-1} u_1,...\varphi^{-1} u_n) \in R_k$. Define an interpretation σ^t such that the interpretation of each O_k under σ^t is $\varphi(O_k)^\sigma$. Then any L-sentence true under the intended interpretation σ is true under the interpretation σ^t—but in the latter case the observable objects of which the sentence is true (for at least one sentence in T) are *different* from the observable objects of which it is true under σ (it is an *unintended* interpretation). How, then, are we to fix reference to things, even observable things, in this way of thinking of reference? It is no good adding more *theory*, for the enriched theory will itself be subject to exactly the same argument in turn.

There are certain similarities to the previous argument, but it is essentially quite different: it appears to be a genuine paradox, with a genuinely skeptical conclusion. For consider a toy example: suppose "the cat is on the mat" is true; then what makes it true may just as well be taken as the fact that the cherry is on the tree or that any member of some highly gerrymandered sets of objects is contained in some other. With that, reference to observable things becomes wholly problematic. With the previous kind of model, call it type (i), those sentences concerning observable entities true under σ_0 not only remain true in the constructed model, they are true of the same members of S. The difficulty is that the theory admits "spurious" models—that models extending the observable can always be constructed, that truth beyond observables comes cheap. In the second kind of model, type (ii), "deviant" models, the constraint on reference to observable things is lost. Arbitrarily different objects can play the role of truth makers, even for statements about observable things.

6. SKEPTICISM ABOUT MEANING

Putnam (1993, 180) viewed the inability to rule out type (ii), deviant models, as genuinely paradoxical—in his words, "as a refutation of any philosophical position that leads to it." His target by this time was not only scientific ("metaphysical") realism but forms of empiricism as well—most notably Quine's. Quine had earlier shown that the same indeterminacy of reference followed from his own verificationist picture of language use and reference. But Quine, unlike Putnam, did not see the conclusion as paradoxical. Rather, it was an extension—and a demonstration of sorts—of his views on underdetermination of meaning in ordinary language. To understand these, we need some more stage setting.

By midcentury it was widely granted that verificationism as a theory of meaning applied not to individual sentences but to collections of sentences or theories as wholes. In "Two Dogmas of Empiricism" (1951) Quine famously drew an unexpected conclusion: that as a result, no verificationist sense could be made of the analytic-synthetic distinction. In *Word and Object* (1960) he went further: not just analyticity or likeness of meaning was without factual content but reference as well. Indeed if words have no meaning over and above what can be determined from observable linguistic behavior, and if such behavior is insufficient to determine the truth conditions of individual sentences as opposed to whole swaths of sentences, then it is futile to ask what words or sentences mean or refer to singly. Reference goes "inscrutable".

Quine (1960) illustrated the point with his thesis of *indeterminacy of translation*: that in translating from one language to another, entirely foreign language—"radical" translation—there could be no matter of fact as to "correctness" of translation of individual sentences. There was only better or worse translation of languages as wholes, more or less adequate with respect to smoothness of discourse.

To this doctrine of indeterminacy only one class of sentences was exempt: "observation sentences," that is, those occasion sentences—sentences dependent on occasion of utterance for their truth value—whose truth can be settled at a glance. Whether composed of one word or many, taken as wholes such sentences can be mastered independently of the rest of language and hence function as the entry point into language. Sentences like "Milk," or "Mama," or "Rabbit" can be learned and produced as observation sentences in appropriate circumstances, whereas taken as terms as they enter into nonobservation sentences, their meanings are indeterminate, admitting multiple translations.

Quine's examples traded on fine distinctions rather than gross trade-offs. Thus "Rabbit" taken as an observation sentence is interchangeable with "Rabbithood again," "Undetached rabbit part," and "Rabbit stage," but now each of these differs when taken as a referring singular term, on entering the "web of belief"—a web underdetermined by overt linguistic behaviour. However, Quine emphasized, radical translation is a one-off event. In practice, for any pair of languages a paradigm of good translation is already to hand, from which any deviation will ipso facto count as bad translation. But in his "Ontological Relativity" (1968) Quine found a new way to illustrate his doctrine. It was the same device as the one used by Putnam some ten years later: take any permutation φ on the universe of discourse of the intended interpretation σ (what Quine called a "proxy function") and use it to generate a deviant model. The result is a shift in ontology: each predicate is reinterpreted as true of the correlates φx of the objects x that it had been true of before. In Quine's (1990, 32) words: "The observation sentences remain associated with the same sensory stimulations as before, and the logical interconnections remain intact. Yet the objects of the theory have been supplanted as drastically as you please."

Quine understood the argument as a demonstration of the "indifference" of ontology, over and above getting all the sentences of a theory (in the logical sense) to come out right, and to that end all that was needed, apart from the connection with stimulus (and thereby verification of observation sentences) was a universe of the right cardinality. This conclusion is obviously reminiscent of the Newman objection and to Russell's form of structuralism, but the moral to draw from it is tempered by Quine's epistemological naturalism. The result is *not* intended—cannot, by Quine's lights, be taken—to undercut scientific knowledge claims as standardly construed. According to Quine (1990, 34), it follows from science itself that it is irrelevant to those claims whether words refer to what we think they do or to their correlates under some proxy function:

> "There's a rabbit" remains keyed to the sensory stimulations by which we learned it, even if we reinterpret the term "rabbit" as denoting cosmic complements or singletons of rabbits. The term does continue to conjure up visions appropriate to the observation sentence through which the term was learned, and so be it; but there is no empirical bar to the reinterpretation. The original sensory associations were indispensable genetically in generating the nodes by which we structure our theory of the world. But all that matters by way of evidence for the theory is the stimulatory basis of the observation sentences plus the structure that the neutral nodes serve to implement. The stimulation remains as rabbity as ever, but the corresponding node or object goes neutral and is up for grabs.

The grounds for the resulting structuralism are naturalistic, or so Quine maintained. And the result is not a structuralist conception of theories in the formal, reconstructed sense: science, for Quine, is to be taken at face value, not in some reconstruction. But it can certainly look like one, or otherwise look self-defeating: for if there is really no matter of fact as to what our words refer to, how *can* we take science at face value? Hence Putnam's complaint. And if the result *is* understood to imply some form of structuralism in the theory-reconstruction sense, then no matter that the theory involved is a first-order theory, the position is no more tenable than Russell's. Putnam's paradox is tolerable but only insofar as the most radical form of meaning skepticism is tolerable, no more and no less.

7. LEWIS'S "RAMSEYAN HUMILITY" AND TWO RESPONSES TO TRIVIALITY

What, then, are we to say? Returning to Putnam's two kinds of models, spurious (type [i]) and deviant (type [ii]), there is the temptation to run them together: both after all will be solved if some constraint not expressible in T can be used to rule out the unwanted interpretations. But what are those constraints? It is natural to insist that the cherry on the tree is not the *cause* of my perception that the cat is on the mat, and that causal relations surely single out the intended interpretation; or at least, predicates of things specified ostensively would be secured (see, e.g., Merrill 1980). But as Putnam points out, that is only to exclude the ostensive act itself from the purview of physicalistic discourse. For if such constraints are themselves formulated and added to the theory T, they will be true in any model of the theory, both spurious and deviant.

Less flat-footed but in the same spirit was David Lewis's response to the problem. Instead of locating the solution in either language users' intentions or their relations to the world, Lewis proposed that we root the solution in the world itself. In "Putnam's Paradox" (1984) Lewis pointed out that the properties and relations on S are gerrymandered ones from the standpoint of the assumed "correct" intended interpretation. So what is needed, he claimed, is a restriction on the sets of things in the world that can serve as the interpretations of the predicates. They must be properties that are especially "eligible": an "elite" class of properties whose extensions are demarcated along lines of "objective sameness and difference." These are the "perfectly natural properties." And whatever the difficulties presented by the latter notion (and of "natural kinds" more generally), it cannot be gainsaid, argued Lewis (1983, 365), that science does deal only with natural properties and relations (and never these gerrymandered ones)—for otherwise swaths of physics practice would simply be unintelligible.

The response feeds into a technical line of reply to Newman's objection that has already been mentioned: if some kind of restriction on the second-order domain of quantification may be motivated, in effect eschewing models that are full (models that contain the power set of the first-order universe) in favor of models (Henkin models) containing only certain "privileged" properties and relations, then the triviality objection against the Ramsey sentence may be blocked. This might explain why Lewis was at such pains to rebut Putnam, as Lewis (1970) had already defended a version of Carnap's structuralism. There he maintained a Carnapian emphasis on holism with respect to the theoretical language: he proposed that the growth of theoretical knowledge, relative to an existing body of theory (which we may consider fully interpreted), is best represented by statements formulated exclusively in logical vocabulary that imply that extant theory. This is essentially to propose a Ramsey sentence with the "observational system" replaced by "old theory", but there was a further difference as well. In accordance with the intuition that genuine knowledge ought to possess an aspect of determinacy, Lewis explicitly rejected Carnap's insistence, already noted, that a quantified second-order variable "does not refer to any particular class." Instead, in Lewis's construction Carnap's second-order existential quantifiers are replaced with iota operators, which demand of any such sentence that it be uniquely satisfied if it is to be true. The advantage (according to Lewis) of such second-order definite descriptions over ordinary first-order ones is that they do not imply of the new vocabulary that there can be more to the meaning of its terms than the connections they make with terms with which we are familiar. So far he is with Carnap. But unlike Carnap, he believes there is indeed a unique set of properties and relations in the world that may or may not be truly described by a final theory.

Our predicament, according to Lewis, is one of "Ramseyan humility." We cannot ourselves discriminate between the various isomorphic relations that could serve as referents of the Ramsey variables, but we must assume (or may at least hope) that there is some unique set that would ground the decision for us, even if we cannot access that ground itself. The perfectly natural properties are precisely this set. Since these are not of our own making, the nontrivial and nonarbitrary nature of our knowledge is secured by the world and its discrimination between isomorphic relations.

Humble or not, it is clearly a form of realism, and it suited empiricists like Bas van Fraassen (2006, 541) to characterize it as such:

> The realist's final gambit comes as no surprise. S/he insists that there is an essentially unique way of "carving nature at the joints," with an objective distinction "in nature" between on the one hand arbitrary or gerrymandered and on the other hand natural divisions of (the physical entity) into a set of parts, and similarly

between arbitrary and natural relations between those parts. This is a postulate (if meaningful at all). What can it possibly mean? The word "natural", its crucial term, derives its meaning solely from the role this postulate plays in completing the realist response [to Putnam's Paradox].

According to van Fraassen, then, the response is ad hoc. To be sure, others have taken Lewis's side: for Robert Stalnaker (2004, 203), "what the account is committed to is only the rejection of an extreme form of nominalism that denies that there is any distinction to be drawn (independently of our theorizing) between more natural and more artificial ways of grouping individuals into classes." But one can hardly deny that we remain largely in the dark as to what can positively be said about natural properties other than that they exist and have the effects we desire of them. And certainly, pace Quine, humility is hardly the right response to type (ii), deviant models.

As a solution to the triviality objections, then, the appeal to perfectly natural properties certainly has its problems. But from a formal point of view it is a solution of sorts.[5] There is, however, a second line of defense that may be attractive to Lewisians.

We saw above that Lewis embraced the Ramsey sentence as a device to make sense of theory change, of introducing new language into old. He was writing, we must suppose, at a time before concerns raised by Thomas Kuhn's account of revolutionary theory change (or at least of referentially incommensurable theory change) had really sunk in. Sentences in the "old" language were not supposed to suffer any change in truth value. As such "old language" corresponded to and replaced the logical empiricists' "observation language." Lewis was quite explicit on this. The old/new distinction he considered much more defensible than the old empirical one (Lewis 2008, 205). Indeed the many (and well-documented) crude and unconvincing aspects of the theory/observation distinction have prompted a number of recent defenses of the Ramsey sentence against the Newman objection. As Joseph Melia

---

5. For more on this see Saunders (2014).

and Juha Saatsi point out, the assumption we made above that the observation and theoretical terms featuring in the original theory have disjoint extensions—do not overlap—turns out to be a crucial assumption of Newman's argument as applied to the Ramsey sentence, at least in the form Jeffrey Ketland (2004) gave it (see Melia and Saatsi 2006 for details on this). For if even one observation predicate can be assumed to share a member with a theoretical predicate, the objection fails to go through. Nor is there any obvious reason to insist on the "no-overlap" principle, from a realist point of view.

It would be ironic if Lewis's alternative distinction between old and new language—better motivated, we may suppose, than the observable/unobservable distinction—were to carry with it the no-overlap principle. But why should it? There is no reason why old and new predicates should not (sometimes) apply to the same things. And if that is the case, then again the troublesome permutations giving rise to the triviality charge may be blocked.

Denial of the no-overlap principle for old and new language thus offers a second line of defense for advocates of the Ramsey sentence. But there remains a problem that is in a sense the mirror image of this solution. While it seems entirely possible that new terms introduced in theory change could sometimes apply to old things, it is just as feasible that such changes may result in old terms coming to acquire extensions different from those that they had before. As already mooted, Lewis's proposal was hardly intended to deal with revolutionary theory change: in particular he shared with Carnap the assumption that old, respectively, observational truths are preserved unchanged. But we know, post-Kuhn, the introduction of new theory can affect language already in place. For example, the term "mass" would have qualified as an "old" term prior to special relativity, but special relativity undoubtedly changed its meaning. It has in effect become an ambiguous term, the use of which is subject to a variety of truth conditions depending on context (Field 1973). Predicates involving mass will apply to the same things, true, when those things have small relative speeds in comparison to the speed of light, but then equally the old predicates involving "mass" must be severely constrained in their use. Thus while Lewis's Ramsey construction, intended as a realist formalization of science at a point in its development, may be defended against the triviality charge, its success *qua formalization* is very much open to doubt.

Relatedly, and as Melia and Saatsi also point out, the very fact that the Ramsey sentence is supposed to capture the growth of theories means that to accompany it with an appeal to "perfectly natural" properties, as Lewis does, is ill-advised for a further reason. For if, as is plausible and as Lewis believes, the perfectly natural properties are the fundamental properties and if fundamental properties are the only admissible truth makers of Ramsey sentences, then it is hard to see how the Ramsey sentence of a theory is any more "open-ended" than the theory TC of which it is the Ramsey sentence. After all, in what sense is R^{TC} better able to characterize discovery and theory change than TC itself, if R^{TC} is interpreted in such a way that it can only ever be false pending the end of inquiry?

But if the Ramsey sentence fares poorly as an expression of the open-endedness of theoretical discourse in Lewis's philosophy, it may do better in others'. Forget about the introduction of new against a backdrop of old terminology; accept that R^{TC} may be true where TC is false just because R^{TC} makes so much weaker a claim than does TC. Make do with substantially weaker knowledge claims than the scientific realist is accustomed to make—claims that are untouched even by the quite massive referential failures that we find in physics. Then the Ramsey sentence of a theory, as a statement that is more nuanced, more structural, and limited in what it claims about the world, may still seem the right thing to say.

8. THEORY CHANGE

The background to this appeal to structuralism is the view that theories in physics have typically failed to refer as a simple matter of historical contingency. The claim was clear in Kuhn, but it found its classic expression in Larry Laudan's "A Confutation of Convergent Realism" (1981) (also known as "the pessimistic metainduction"). Every "best" theory of the past seems to be have been falsified not only in the sense that its predictions turned out to be merely approximately true (if even that), but also in the sense that the central explanatory terms of each theory failed to refer or to refer determinately. Given the assumption that we are not now living in a specially privileged time, terms in our current favorite theories will equally, no doubt, fail to refer as well.

The argument is of course Poincaré's. It was Worrall who, in his paper "Structural Realism: The Best of Both Worlds?" (1989), rehearsed Poincaré's conclusion: that what continuity there is resides in form or structure rather than the manifest content of a physical theory and that this was enough to do justice to the realist's conclusion that *something* must be right about scientific theories if their success is to be explicable. Thus while the facts of theory change undermined Lewis's version, a form of structuralism can yet be motivated by those same facts through its offer of "the best of both worlds."

Worrall's account of the form/content distinction was, however, if anything even more meagre than Poincaré's and certainly less clear than Russell's: he gave few clues as to what the appropriate characterization of structure was beyond likeness of dynamical equations. His example, following Poincaré, was the equations putatively describing the motion of the ether in transmitting electromagnetic waves, equations that were retained by James Clark Maxwell, and even by Albert Einstein, let the ether fall where it may. But the approximate empirical accuracy of certain equations either side of theory change was never by itself in question. And without some firm criteria as to what exactly "structure" excludes, the position is in danger of collapsing into banality. Worrall himself admitted that the appeal to "form" was only a starting point as a developed realist position applicable to all of (at least) mathematical physics; that something both considerably sharper and more general was needed. After all, as Stathis Psillos argued, the course of history equally bears out the continuity of "natures" or "content", unless these are made implausibly narrow (and "structure" correspondingly broad). In Psillos's words, "there is no sense in which Fresnel was 'just' right about the structure of light-propagation and wrong about the nature of light, unless of course one understands 'structure' so broadly as to include the principle of the conservation of energy and the theoretical mechanism of light-propagation" (Psillos 1999, 159).

It was in the face of this and allied criticism that Worrall fell back on the Ramsey sentence. The form/content distinction was to be made out as the distinction between the Ramsey sentence of a theory TC (or just T, coordinating rules C no longer being in favor) and the theory itself (Worrall and Zahar 2001). The distinction was moreover drawn epistemically, as with Russell and Carnap: "Structural realism is in effect committed to the view that the cognitive content of a theory is fully captured by its Ramsey sentence" (Worrall and Zahar 2001, 236). And clearly, in the context Worrall is responding to—namely, that of theory change—the vocabulary cannot be divided into theoretical terms and "old" terms, for it is the structural continuities between the theoretical terms of old and new theories that constitutes Worrall's whole point. Hence Worrall and Elie Zahar assume the vocabulary is divided along the same lines as Carnap and Hempel did.

However, the very case that Worrall discussed at any length gives the lie to the idea that the Ramsey sentence can be usefully enlisted to capture continuity through radical ontological loss. The reason is that any loss at the ontological (interpreted, first-order) level, unless it played strictly no role in the deduction of observational consequences, is likely to have problematic repercussions at the structural level as defined by the Ramsey sentence. Thus consider any conceivable Ramsey expression of "the theory of electromagnetism" before and after the emergence of the electromagnetic field. The loss of the ether entailed the loss of the only viable candidate for an absolute frame of reference; giving up absolute simultaneity, one supplants Newtonian by relativistic dynamics. Thus let Newtonian principles play ever so small a role in the deduction of electromagnetic phenomena from Maxwellian principles: then the two Ramsey sentences, pre- and postrelativity, will be logically and formally distinct. Who is to say that the latter is a conservative extension of the former rather than its wholesale replacement? By what yardstick is this difference to be measured?

The bottom line is that there is in general no reason to believe that Ramsey sentences will be preserved exactly either side of theory change, and the syntactic framework in which it is couched offers us no resources for capturing any approximating relationships that may nevertheless exist between successor theories. Thus while it remains that some sort of structuralism seems our best hope in the face of theory change, a Ramsey-style expression of it is likely to prove a dead end.

9. THE SEMANTIC APPROACH

Common to these expressions of structuralism found in Carnap, Putnam, and Lewis is the use of syntactical methods in the definition of theories. The semantic conception of theories is an explicit attempt to get away from all that, an attempt to formulate and interpret theories not in terms of syntactical structure but in terms of models.

The approach goes back to E. W. Beth's (1949) attempt at logical analysis of the actual content of the sciences and Patrick Suppes's (1961) appeal to "actual mathematics" rather than logic. His "A Comparison of the Meaning and Use of Models in Mathematics and the Empirical Sciences" did as the title said, concluding: "The concept of model in the sense of Tarski may be used without distortion and as a fundamental concept in all of the disciplines [considered].... In this sense I would assert that the meaning of the concept of model is the same in mathematics and the empirical sciences." (Suppes (1961, 165). This paved the way for the attempts by Joseph Sneed and Wolfgang Stegmüller and by Suppes himself for formalizations of theories in set theory. Van Fraassen, in contrast, took up Beth's approach but to a similar end. The semantic approach was also important to his later development of "constructive empiricism" and to Ronald Giere's related project of "constructive realism." James Ladyman (1998) likewise saw in the semantic approach an important point of difference from realism as conventionally phrased and used it as the basis for his favored "ontic" version of structural realism. According to this, structuralism reflected not so much an epistemic constraint as simply the way the world actually is.

Common to these views was the idea that theories should be understood as extralinguistic structures, not as systems of statements—that in fact the main mistake of logical empiricism was its emphasis on language:

> Perhaps the worst consequence of the syntactic approach was the way it focused attention on philosophically irrelevant technical questions. It is hard not to conclude that those discussions

of axiomatizability in restricted vocabularies, "theoretical terms", Craig's theorem, "reduction sentences", "empirical languages", Ramsey and Carnap sentences, were one and all off the mark—solutions to purely self-generated problems, and philosophically irrelevant. The main lesson of twentieth-century philosophy of science may well be this: no concept which is essentially language-dependent has any philosophical importance at all. (van Fraassen 1980, 56)

Models, in the Tarskian sense, are set-theoretic structures. Set theory, we may grant, is extralinguistic, but it is logic in our general sense. And whatever the virtues of models or set-theoretic constructions over formal axiomatic systems, they still involve formal reconstructions of scientific theories in terms other than those in which they are found—or actually used. For all that the semantic approach to theories is supposed to involve mathematical rather than linguistic representation, it singles out a branch of mathematics—set theory—that has few if any scientific applications. Physical theories are stated and applied in terms of algebra, functional analysis, differential geometry, and group theory. In fact few of Suppes's own case studies were drawn from physics. And what, to take just one example, has Niels Bohr's model of the atom to do with a model in the Tarskian sense?

The appeal of the semantic approach to models as used in actual science is, in most cases, simply false advertising. Meanwhile its distancing from formal linguistics creates problems of its own. In casting theories into systems of statements, the formal, axiomatic approach at least aspired to *say* something about the world, even if trivialized (the Ramsey objection, Putnam type [i] models) or deviant (Putnam type [ii] models, Quine's ontological relativity). How, if theories are families of set-theoretic structures, do *they* say anything about the world? Anjan Chakravartty (2001, 331) has pushed the problem in the following terms:

> A model can tell us about the nature of reality only if we are willing to assert that some aspect(s) of the model has a counterpart

in reality. That is, if one wishes to be realist, some sort of explicit statement asserting a correspondence between a description of some aspect of a model and the world is inescapable. This requires the deployment of linguistic formulations, and interpreting these formulations in such a way as to understand what models are telling us about the world is the unavoidable cost of realism. Scientific realism cannot be entertained unless we are willing to associate models with linguistic expressions (such as mathematical formulae) and interpret such expressions in terms of correspondence with the world. In the absence of this kind of assertion, there is no realism. Theories can't tell us anything substantive about the world unless they employ a language.

This is surely a problem for realists; however, it is also a problem for antirealists who, like van Fraassen, are wedded to the semantic approach, for they too must refer at least to observable structures. How are they to do this, precisely?

Ladyman's formulation of structural realism ("ontic" structural realism) is close to van Fraassen's empiricism: the central difference is supposed to be that structures, whether in the world or as abstract entities, carry modal import (Ladyman 1998). In *Everything Must Go* (2007) Ladyman and Don Ross invoke Carnap's distinction between the "formal" and the "material" mode. They frame their position thus:

> We will later say that what exists are ("real") patterns. "Real patterns" should be understood in the material mode. Then "structures" are to be understood as mathematical models—sometimes constructed by axiomatized theories, sometimes represented in set theory—that elicit thinking in the formal mode. We will argue that there is no tension between this view of structures, and the claim that "the world has modal structure". Objective modalities in the material mode are represented by logical and mathematical modalities in the formal mode. (Ladyman and Ross 2007, 119)

Later they make clear their sympathies for the view that the distinction between abstract and concrete structures is not clear-cut (Ladyman and Ross 2007, 159–161); reference is to be made out either by virtue of an identity of the two or by talk—in the material mode—of some fragments of abstract structures as real. They evidently embrace a deflationary view of the role of language in mediating reference, as did van Fraassen in *The Scientific Image* (1980).

More recently, however, van Fraassen has recognized that the puzzles of reference and triviality are not so easily laid to rest (see in this respect Demopoulos 2003). Indeed whether because the semantic view does not wholly rule out syntactic methods (Ladyman and Ross allow that abstract structures are "sometimes constructed by axiomatic theories" (2007, 119)) or because these puzzles arise as much with set-theoretic representations as with those mediated by formal syntax (Putnam's "Models and Reality" was well named), he grants that they afflict scientific representation quite generally.[6] In his book *Scientific Representation* van Fraassen (2008, 3) proposes to solve those puzzles, what he there calls "paradoxes of perspective," by use of intentional and indexical idioms:

> I will begin with protagonists in the historical *Bildtheorie* debate, and show how it was already implicitly challenged by an apparent paradox at the very heart of its conception of science. I will then take the message about the role of indexicality in scientific representation to the paradoxes that beset, bedeviled, or otherwise preoccupied Bertrand Russell, Rudolf Carnap, Hilary Putnam, and David Lewis. For my part I will propose an empiricist structuralism, in contrast to structural realism, as a view that can stand up to these challenges.

With that we come full circle. Appeals to indexicality (intuition) and intentionality (mind dependence) were precisely what the neo-Kantians took themselves to eliminate from Kant's philosophy to secure an account of representation and objectivity.

---

6. This point, as well as others in this section, are discussed in greater detail in Lutz (2014).

10. THE NATURALIST STANCE

It is time to take stock. We have seen the origins of structuralism in neo-Kantianism, but the question of the correspondence between mental representations and the external world to be answered by a purely philosophical theory is nowadays of interest to no one—van Fraassen is right about that. The question makes sense—is forced even—only from a naturalist perspective, using science to see how the trick is done. The appeal to "perfectly natural properties" in an attempt to ward off the subsequent triviality objection is, from a naturalist perspective, simply the reminder that we are using science to articulate our knowledge of that same science—but using it to delineate properties in the world, not provide a philosophical theory of how words latch onto things.

We saw further that Melia and Saatsi argue that the Newman objection to epistemic structuralism is best circumvented not by appealing to elite properties but by appealing to "mixed" predicates—predicates that apply to both observable and unobservable objects or to the extensions of both old and new terms. Such a move deflects Newman's criticism as applied to the Ramsey sentence, which Worrall took to be the appropriate vehicle for expressing realism in the wake of revolutions. But structure thus construed does not in fact solve any of the real problems of theory change, still requires an untenable form of the theory/observation distinction, and assumes the availability of reconstructions in set theory and logic of scientific theories that are in practice unobtainable.

The semantic approach to theories, like Worrall's original appeal to structural rather than ontological continuity, promised to return to the mathematics deployed in real-life scientific theories but—notable exceptions to one side—appealed to the same formal logical methods, whether set-theoretic or syntactical, and led to the same logical puzzles and to a similar problem of theory reconstruction as bedevils the axiomatic approach. Motivated as a solution to these puzzles, it has therefore disappointed. Motivated by a diagnosis of the ills of logical empiricism as based on linguistic analysis, it is in need of an analysis of language use all its own—or else of a picture of the world as itself set-theoretic, a carbon copy of the mathematical abstractions in which physical theories are to be newly defined.

STRUCTURE AND LOGIC

This discussion has shown that the thicket of problems in which structuralism is enmeshed is a dense and confusing one. But what is clearly at issue is a plethora of quite different motivations for structuralism, many of which, from a naturalist perspective, are moribund. We have encountered:

> Motive (1): neo-Kantianism. The external world can be nothing like sensation; the two can be akin only with regard to structure. (This applies to everyday objects too.)
> Motive (2): Ramseyan humility. The true nature of unobservable entities cannot be known but only their structure and relations with observable things.
> Motive (3): the problem of meaning. Theories are systems of sentences, and the meanings of theoretical terms is partly conventional, partly factual. Their factual content is structural.
> Motive (4): theories are extralinguistic. As such they should not be characterized in syntactical terms, they should be defined as families of models or "structures."
> Motive (5): the pessimistic metainduction. Physical theories show no continuity at the level of unobservable things, properties and relations in the world; the only continuity possible is in terms of their "structure."

To which we may add, as common background to all physically minded philosophers and physicists:

> Motive (6): modern physics. According to physical theories, and especially field theories and quantum mechanics, the world cannot be understood as comprised of things in the ordinary logical sense; the world is made of patterns of forms or "structures."

Motives (1)–(3) are broadly foundationalist in the epistemological sense; (5) and (6) and possibly (4) are available to naturalists. All invite or require an account of "structure." Motive (3) invites an account in terms of syntactical structure, (4) in terms of set theory. As an a priori doctrine, (1) is likewise linked to logic and set theory. Motive (2) is in a certain sense a platitude: surely unobservable entities cannot be known in the same way as observable ones. In another sense, with the emphasis on truth, it is rejected by naturalists: the "true" nature of tables and chairs is not known by direct observation either. As naturalists rejecting, with Quine, the conventionalist's project and hence motive (3), only (5) and (6) and possibly (4) still have currency.

Our concern is therefore with (4)–(6). Given the persistence of the "paradoxes of perspective," as van Fraassen calls them, is it the structuralism itself that is the root difficulty or the formal logical methods underpinning it? The answer is both: the problem lies with structuralism, because no other idiom but syntax and set theory has been found to define it, and the problem lies with formal logic and set theory, because the puzzles we have encountered arise independently of embracing structuralism as a metaperspective on scientific theories.

That raises the hope that only solve those logical puzzles—whether by appeal to a purely philosophical theory of indexicals or by appeal to more sophisticated formal, logical methods—and the problems of structuralism will be solved as well. But the first avenue is hardly available to epistemological naturalists, while the second has the feel of being in a hole and digging deeper. Both may seem to be solutions to "purely self-generated problems," for they both involve the wholesale reconstruction of physical theories. The entire project rests on make-believe: we do not have such reconstructions, whether in formal axiom systems or as a family of models in set theory. And even if we did, we have as yet no understanding of how the structuralist's objective in showing continuity through theory change is to be achieved. By what metric are models in set theory or formal axiomatic systems to be compared with one another? The appeal to structure, in the formal logical and set-theoretic sense, provides a definite notion of form but at the price of a sufficiently flexible notion of similarity of form. If our motive is (5), we seem no closer to our goal.

We have said that structuralism has found no other expression save in logic and set theory. Is that simply a historical fact, and might one hope to do better? What *is* clear is that insofar as the motive of structuralism is (5), more analytic resources are required than those designed to apply only to a single theory considered in isolation: an overarching framework in which intertheoretical relationships can be comprehensively surveyed is what is needed. But now there is a problem, for in constructing this framework any method that assumes that a formalization of theory is already available will be of little use, for what formalizations are appropriate is surely a large part of what is in question. The answer to this conundrum is to appeal to an overarching framework that is truly universal, topic-neutral, and of unrestricted scope. But it seems that the only such framework is logic, or set theory, or some combination of both.

However, there is another answer to our conundrum and as far as we can see the only one:[7] *consider the physical theories to be compared as they are actually presented, using the mathematics in which they are ordinarily defined.* Therein lies the solution to the problem set by Laudan's so-called pessimistic metainduction if it is to be found anywhere, and there, whether trivially or implicitly, lies the solution to the problem of theory formulation—for either no reconstruction will be needed, or the comparison will require reformulations specific to the two theories involved. As for motive (6) and intimations of structuralism internal to physics, they will pan out as they may. If there is common ground to intertheory relations, we will learn it in whatever shared idiom there is to be found. But with that there is a broader conclusion to draw: it is that *there is no metaperspective available, no overarching formal system in which intertheoretical relationships can be classified and displayed other than that provided by the sciences themselves.* Philosophy of science, on our proposal, gives way to philosophy of physics and philosophy of the special sciences.

Our argument evidently is similar to Quine's (1969) concerning the point, or lack of it, of rational reconstruction in epistemology. As with Quine's proposal of a "naturalized" epistemology, it has an air of defeatism—of giving up on distinctively philosophical problems and methods.

---

7. It is possible that category theory might provide an alternative, topic-neutral foundation to mathematics. But that it will also turn out to provide a foundational theory of reference, and an appropriate framework for an account of theory structure in the physical sciences, seems more like wishful thinking.

But look closer and the problems remain: just as with the semantic approach, there is still the question of reference. Pace van Fraassen, there is still the nature of linguistic expression, of how to talk about physical things in the light of theory, and if not in simple, declarative sentences, the bedrock of elementary logic, then how else are we to do so? And here formal methods—regimentation of language—may be an aid to clarity, for settling questions of identity, for example, in the individuation of things. Quine's (1948) dictum, "to be is to be the value of a bound variable" has the same force that it ever did in determining someone's ontological commitments in the light of what he or she says. But do not Quine's ontological relativity and Putnam's paradox follow in train?

So they do—but now, looked on squarely from our newfound physicalistic perspective, they take on a different character. For at bottom they concern the relationship between abstract set-theoretic structures—models in Alfred Tarski's sense—and physical things. Why would one think such a representation is any easier to make sense of, is any more perspicuous or direct, than a representation of reality in terms of a coordinate system in a physical theory like special or general relativity? In *one* way of course it is quite simple—the way that implicitly or explicitly relies on intentional idioms or indexical reference: coordinates mean precisely what we intend them to mean, no more and no less, and the same can be said of names, predicates, and values of variables in formal logic. But other ways were also found to make sense of coordinates in space-time theories—*had* indeed to be developed to give empirical content to the notion of inertial frame and to Newton's notion of absolute time, and those other ways made do with qualitative, even (from a mathematical point of view) *structural* notions, free (or at least apparently free) from intentional and indexical idioms.

The connections between the use of coordinates in physics and Putnam's paradox have been made out before,[8] but there is an even closer analog—namely, the parallel with particle labels in classical and quantum statistics. There the relevant symmetry is the permutation group—exactly the symmetry group underlying Putnam's type (ii)

---

8. See, e.g., Liu 1996; Rynasiewicz 1996; Saunders 2003; Stachel 2002, 2005. The link with symmetries more generally was first pointed out by Tim Maudlin (1988).

models and Quine's proxy functions. It is true that the "neutral nodes" encountered in statistical mechanics are far from observable, and as such their permutability—yielding redescriptions of the *same* physical states of affairs—is free of any hint of paradox. But among the tasks carried through in that domain is an account of how an underlying symmetry at the microscopic level is consistent with a lack of symmetry at the macroscopic—in the passage from indistinguishable particles to distinguishable ones. That story, however, and in keeping with our main conclusion, is not easily paraphrased.[9]

Acknowledgment

With thanks to Sebastian Lutz for comments and constructive criticism.

References

Beth, E. W. 1949. "Towards an Up-to-date Philosophy of the Natural Sciences." *Methodos* 1: 178–185.
Carnap, Rudolf. 1956. "The Methodological Character of Theoretical Concepts." *Minnesota Studies in the Philosophy of Science* 1:38–76.
Carnap, Rudolph. 1963. 'Replies and Systematic Expositions.' In the *Philosophy of Rudolph Carnap*, ed. P Schlipp, La Salle: Open Court.
———. 1966. *Philosophical Foundations of Physics*. Edited by Martin Gardner. New York: Basic Books.
Chakravartty, Anjan. 2001. "The Semantic or Model-Theoretic View of Theories and Scientific Realism." *Synthese* 127 (3): 325–345.
Demopoulos, William. 2003. "On the Rational Reconstruction of Our Theoretical Knowledge." *British Journal for the Philosophy of Science* 54 (3): 371–403.
Field, Hartry. 1973. "Theory Change and the Indeterminacy of Reference." *Journal of Philosophy* 70 (14): 462–481.
Hempel, Carl G. 1958. "The Theoretician's Dilemma." In *Concepts, Theories, and the Mind-Body Problem*, ed. Herbert Feigl, Michael Scriven, and Grover Maxwell, 37–98. Minnesota Studies in the Philosophy of Science 2. Minneapolis: University of Minnesota Press.

---

9. See Saunders 2006 and, in more detail 2013, 2014.

Ketland, Jeffrey. 2004. "Empirical Adequacy and Ramsification." *British Journal for the Philosophy of Science* 55 (2): 287–300.

Ladyman, James. 1998. "What is Structural Realism?" *Studies in the History and Philosophy of Science* 29 (3): 409–424.

Ladyman, James, and Don Ross. 2007. *Everything Must Go: Metaphysics Naturalised*. Oxford: Oxford University Press.

Laudan, Larry. 1981. "A Confutation of Convergent Realism." *Philosophy of Science* 48 (1): 19–49.

Lewis, David. 1970. "How to Define Theoretical Terms." *Journal of Philosophy* 67 (13): 427–446.

———. 1983. "New Work for a Theory of Universals." *Australasian Journal of Philosophy* 61 (4): 365.

———. 1984. "Putnam's Paradox. *Australasian Journal of Philosophy* 62 (3): 221–236.

———. 2008. 'Ramseyan Humility'. In *Conceptual Analysis and Philosophical Method*, eds. David Braddon-Mitchell and Robert Nola, pp. 203-222. Cambridge, MA: MIT Press.

Liu, Chang. 1997. "Realism and Spacetime: Of Arguments against Metaphysical Realism and Manifold Realism." *Philosophia Naturalis* 33:243–263.

Lutz, Sebastian. 2014. "What's Right with a Syntactic Approach to Theories and Models?" *Erkenntnis* 1-18,

Maudlin, Tim. 1988. "The Essence of Space-time." *PSA: Proceedings of the Biennial Meeting of the Philosophy of Science Association* 1988 (2): 82–91.

Melia, Joseph, and Juha Saatsi. 2006. "Ramseyfication and Theoretical Content." *British Journal for the Philosophy of Science* 57 (3): 561–585.

Merrill, G. H. 1980. "The Model-Theoretic Argument against Realism." *Philosophy of Science* 47 (1): 69–81.

Newman, M. H. A. 1928. "Mr. Russell's 'Causal Theory of Perception.'" *Mind*, n.s., 37 (146): 137–148.

Psillos, Stathis. 1999. *Scientific Realism: How Science Tracks Truth*. London: Routledge.

Putnam, Hilary. 1980. "Models and Reality." *Journal of Symbolic Logic* 45 (3): 464–482.

———. 1981. *Reason, Truth, and History*. Cambridge: Cambridge University Press.

———. 1985. *Realism and Reason*. Vol. 3 of *Philosophical Papers*. Cambridge: Cambridge University Press.

———. 1993. "Realism without Absolutes." *International Journal of Philosophical Studies* 1 (2): 179–192.

Quine, W. V. 1948. "On What There Is." *Review of Metaphysics* 2: 21–38.

———. 1951. "Two Dogmas of Empiricism." *Philosophical Review* 60:20–43.

———. 1960. *Word and Object*. Cambridge, MA: MIT Press.

———. 1968. "Ontological Relativity." *Journal of Philosophy* 65 (7): 185–212.
———. 1969. "Epistemology Naturalized." In *Ontological Relativity and Other Essays*, 69–90. New York: Columbia University Press.
———. 1990. *Pursuit of Truth*. Cambridge, MA: Harvard University Press.
Ramsey, F. P. 1931. *The Foundations of Mathematics and Other Logical Essays*. Edited by R. B. Braithwaite. London: Kegan Paul, Trench, Trubner.
Russell, Bertrand. 1927. *The Analysis of Matter*. London: George Allen and Unwin.
Rynasiewicz, Robert. 1996. "Is There a Syntactic Solution to the Hole Problem?" *Philosophy of Science* 63 (supp.): S55–S62.
Saunders, Simon. 2003. "Indiscernibles, General Covariance, and Other Symmetries: The Case for Non-reductive Relationalism." In *Revisiting the Foundations of Relativistic Physics: Festschrift in Honor of John Stachel*, ed. Abhay Ashtekar, Robert S. Cohen, Don Howard, Jürgen Renn, Sahotra Sarkar, and Abner Shimony, 151–174. Dordrecht, Netherlands: Kluwer Academic.
———. 2006. "On the Explanation for Quantum Statistics." *Studies in the History and Philosophy of Modern Physics* 37 (1): 192–211.
———. 2013. "Indistinguishability." In *The Oxford Handbook of Philosophy of Physics*, ed. Robert Batterman, 340–380. New York: Oxford University Press.
———. 2014. "On the Emergence of Individuals in Physics." In *Individuals across the Sciences*, ed. Alexandre Guay and Thomas Pradeau. Oxford: Oxford University Press.
Stachel, John. 2002. "'The Relations between Things' versus 'the Things between Relations': The Deeper Meaning of the Hole Argument." In *Reading Natural Philosophy: Essays in the History and Philosophy of Science and Mathematics*, ed. David B. Malament, 231–266. Chicago: Open Court.
———. 2005. "Structural Realism and Contextual Individuality." In *Hilary Putnam*, ed. Yemin Ben-Menahem, 203–219. Cambridge: Cambridge University Press.
Stalnaker, Robert. 2004. "Lewis and Intentionality." *Australasian Journal of Philosophy*, 82 (1), 199-212.
Suppe, Frederick. 1977. *The Structure of Scientific Theories*. Urbana: University of Illinois Press.
Suppes, Patrick. 1960. "A Comparison of the Meaning and Use of Models in Mathematics and the Empirical Sciences." *Synthese* 12 (2–3): 287–301.
van Fraassen, Bas. 1980. *The Scientific Image*. Oxford: Clarendon.
———. 2006. "Representation: The Problem for Structuralism." *Philosophy of Science* 73 (5): 536–547.
———. 2008. *Scientific Representation: Paradoxes of Perspective*. Oxford: Clarendon.

Worrall, John. 1989. "Structural Realism: The Best of Both Worlds?" *Dialectica* 43 (1–2): 99–124.

Worrall, J., and E. Zahar. 2001. Ramseyfication and Structural Realism. Appendix IV in Zahar E. *Poincaré's Philosophy: From Conventionalism to Phenomenology*. Chicago and La Salle (IL): Open Court.

Chapter 6

Evolution and Revolution in Science

JARRETT LEPLIN

1. THE COMPLEXITY OF SCIENTIFIC CHANGE

At present the doubling time for scientific knowledge is under fifteen years and decreasing. Knowledge of the world is increasing at an accelerating rate. But contrary to the popular picture of cumulative growth, science historically has also changed by rejecting established theories. Much once thought to be known is now thought to be incorrect and so never to have been known after all. Changes in science both add and detract. There seem to be both continuity and disruption. It is therefore possible, if perverse, to deny that scientific knowledge actually *grows* at all. Maybe what happens is that the scientific outlooks of different epochs are just *different*, none epistemically better off than others.

An evolutionary picture of scientific change suggests continuous growth. In as much as we regard evolutionary change as developmentally advantageous, this picture underwrites the received view that scientific change is progressive. A revolutionary picture, by contrast, suggests that change consists in disruptive substitution of radically divergent perspectives that resist epistemic comparison. We shall consider models of each type of change and assess their applicability to science.

2. DARWINIAN EVOLUTIONARY MODELS

We are accustomed to thinking of evolutionary change in terms of the development of living species in accordance with Darwinian principles of natural selection. An evolutionary model of change based on these principles includes the following requirements:

1. *Diversity.* The generation of diversity of traits or properties for the organisms whose change is modeled; over time the organisms exhibit great variation.
2. *Adaptation.* Differential adaptiveness of properties; properties differ to varying degrees in the extent to which they help or hinder the survival of their possessors, which accordingly are more or less *fit* depending on which properties they happen to possess.
3. *Competition.* Resources are insufficient to sustain all the organisms, so that those with more adaptive properties are more likely to survive than others.
4. *Inheritance.* Properties that affect survival are inherited through reproduction.

In *On the Origin of Species* (1964) Charles Darwin argued that the characteristics and geographical distribution of life forms found on earth are fully explainable on the basis of these requirements. These requirements account for all the evidence available to the naturalist without the introduction of any additional hypothesis of design or goals in nature. In particular the variation of properties is generated randomly with respect to their affect on the fitness of the organism. Can these requirements do the same for science? Can we explain the development of scientific theories and hypotheses, of what we normally take to be scientific knowledge, in terms of the satisfaction, by science, of the requirements of the evolutionary model?

As scientists are themselves products of biological evolution, one might regard their properties, including their beliefs, ideas, and behaviors, as products of evolution. Cognition then arises through some biological process that generates great variety, and different cognitive states differentially affect the survivability of the cognizer. States that impart a competitive advantage will more frequently be transmitted to further generations of cognizers.[1] For the mechanisms of transmission we can look to the social institutions of science, such as instruction and employment. What we consider science, or more broadly, knowledge, consists of whatever cognitive states have been selected for and retained through generations of cognizers.

It does not seem plausible, however, to suppose variation in cognitive states to arise with the spontaneity and randomness of the physical characteristics of organisms that affect the course of biological evolution. Scientific ideas result from intellectual deliberation in the service of epistemic ends. Whereas genetic mutation has no connection with the advantageousness of the new traits it produces, new cognitive states arise in the attempt to solve specific problems and achieve specific goals. It does not seem possible to understand the generation of cognitive states without reference to design and purpose. Some philosophical accounts of scientific method leave the generation of hypotheses to unreconstructable insight or accident, restricting scientific reasoning to the evaluation of ideas already proposed. But these accounts are legacies of a positivistic tradition that has been discredited as unhistorical and dogmatically embedded in an empiricism too rigid to make sense of the semantics of scientific discourse.

Perhaps modes of intellectual deliberation, rather than the cognitive states they generate, can be the units of natural selection. Perhaps the way we think has an evolutionary explanation, even if what we think does not. But if, in learning about the world, we learn how to learn about the world and so improve our methods of inquiry on much the same basis that we improve our empirical beliefs, the implausibility of supposing that science results from biological selection carries over. Scientific methods are skillful, and skills seem not to be randomly generated but developed and perfected in the service of goals.

A more important objection to applying the evolutionary model to scientific knowledge is that we distinguish the evolutionary

1. See Ruse 1986.

advantageousness of cognitive states from their epistemic value. There seems to be no natural connection between whether a belief is true or justified by evidence and whether believing it has survival value. At least any such connection would seem to be limited to immediate cognitive responses to the environment and not to extend to science. In fact the direction of theorizing in fundamental science may have no conceivable bearing on any pragmatic interest. It is difficult to square the thesis that knowledge evolves biologically with the conceptual independence of epistemic and evolutionary value. What would be the evolutionary value of this independence?

Ultimately, an evolutionary picture of scientific growth is forced to abandon the connection of knowledge with truth. Knowledge becomes a way of or strategy for interacting with the world. Different strategies might be compared as to effectiveness, but to benefit by such a comparison is not necessarily to understand the world correctly or even better in any epistemic respect. This disconnection between science and truth is completely unacceptable. Although science serves other values than the acquisition of truths, there can be no serious question that progress in science advances the goal of a deeper and more accurate understanding of the nature of the world at levels inaccessible in ordinary experience. Scientific progress is not simply a matter of improving evolutionary fitness.

Instead of trying to subsume science under biological evolution, we can try to find at work in science processes that instantiate more general evolutionary principles. The most obvious Darwinian feature of science is competitiveness. Scientific ideas are subjected to severe test and do not survive without passing. There is a scarcity of resources in the form of limitations of funding, faculty positions, equipment, computer time. But the clearest source of competition is simply the fact that science generates rivalries; incompatible rival hypotheses are proposed to explain the same evidence, and the process of testing promotes consistency through elimination. Theories compete not just as animals compete for food but also as they compete for mates. Whether or not a less adaptive theory suffers from lack of support, it will eventually perish by the loss of position to a rival. The evolutionary effects of competition seem to account well for much of scientific development.[2]

In other respects the evolutionary model seems less satisfactory. What exactly does fitness in science consist in? What is the analog of strength, speed, or guile in the animal? Theories survive by passing tests, but what properties enable them to do this? If the answer is the accuracy of their account of the world, the correctness of their hypotheses as to the experientially inaccessible mechanisms that produce the empirical results by which they are tested, then the evolutionary model would seem to be incidental. It tells us that the reason for our current perspective is that other perspectives have failed, but it does not tell us why they failed or why any theory succeeds. This information must be imported from outside the model. The informative answer to why and how science changes is that science learns more of the truth; evolutionary selection is merely the result of this epistemic achievement.

For evolution to be a satisfactory explanation of scientific change, the adaptiveness of theories, their ability to pass tests, must be arbitrary or accidental in the way that certain properties of an organism just happen to contribute to fitness in its environment. In some other environment the same property could be disadvantageous. This does not seem to be true of science. We take the empirical success of theories to be a measure of their service to our epistemic ends. If scientific change is evolutionary, we are wrong to do so. There is no independent goal for evolutionary success to serve.

3. GENERAL EVOLUTIONARY MODELS

It would be premature to deny, for this reason, that science evolves. For Darwinian evolution is not the only kind. If it were, one might well wonder why Darwin's theory of evolution should have been so called to begin with.

---

2. According to Bas van Fraassen (1980), theory choice in science is entirely a process of evolutionary competition.

The original meaning of "evolution" is the unfolding or extension of that which has been wrapped up or enveloped. Evolution is a process whereby the concealed properties or nature of a thing are revealed or, by extrapolation, whereby the potential nature of a thing is realized. The early uses of the term in biological literature (from 1762) speak of the evolution of the entire or complete state of an animal or plant from its egg or seed.[3] More generally, evolution is the development of organisms from their rudimentary to their mature states.

Ironically, this basic idea is at odds with Darwin's preemption of design from biology. Some design or plan is present at least implicitly at the beginning; evolution realizes this plan by bringing to fruition the capacities or potentials with which the evolving entity is originally endowed. Perhaps this inconsistency in the concept of evolution contributes to the confusion over the compatibility of evolutionary biology with divine creation.

What is important in the core concept of evolution, common to pre- and post-Darwinian uses of the concept, is that change occurs in accordance with general laws, rules, or principles that explain its course. It is characteristic of pre-Darwinian uses that the result of evolutionary change is foreshadowed or prefigured, if not foreordained, in the original nature of the evolving entity, and what is required to reach the result is essentially that the evolutionary process be uninterrupted, that the entity be allowed to develop naturally, meaning without interference. There must be a way that it is natural for the entity to develop, a pattern of progress suited to it. This pattern can be captured in laws or principles that explain, given the entity's original nature, the properties it comes to have. The particular principles of the Darwinian model are incidental. They govern the evolution of biological species but are not necessarily the right principles to apply to the evolution of science. Scientific change is evolutionary in the core sense if the development of science follows general principles that explain its later stages on the basis of earlier stages. Evolutionary epistemology is overly and unnecessarily restricted if it relies on the Darwinian model.

---

3. *Oxford English Dictionary*, compact ed., s.v. "evolution."

The reason to tolerate this restrictiveness is that other evolutionary models for science are difficult to square with the data of scientific practice. Perhaps the most striking feature of science is the value to it of unanticipated originality. An appeal to originality may be the best we can do by way of supplying the diversity required in the Darwinian model. We do not really know what the source of diversity or the mechanism of diversification is in science. Of course neither did Darwin know what it is in biology, but at least he was proved right about there being one and about its operating according to natural law. By contrast, there does not seem to be any reason to expect that the process that supplies science with a rich influx of new and different ideas will be subsumable under general principles that explain the course that science follows.

Nor is science free from influence by outside agency, able autonomously to evolve in accordance with whatever laws or principles might be imagined to direct its path. There is a strong tradition of autonomy in science, of internal standards and disinterested pursuit of goals set from within. But science is highly responsive to social forces that set priorities by allocating resources. And it is even more subject to influence by technological changes that from the perspective of internal directions of research often appear adventitious or serendipitous. It does not look like there are any very significant laws for science to evolve by, nor would science be free to follow them unimpeded if there were. Thus the prospects for a non-Darwinian evolutionary model of science are not promising.

I have been deliberately unspecific in speaking of laws, rules, and principles of change, because it is disputatious what exactly is required for something to be a law and unclear whether laws are required for explanation or prediction. There are certainly facts as to how science has changed, and the question is what conditions these facts need to satisfy for this change to count as evolutionary, broadly speaking. Perhaps it is too restrictive to require that the facts instantiate laws. But it is too permissive to be satisfied with mere patterns or regularities that there is no good reason to expect future science to follow or good reason to think that past science would have followed had contingencies been different. Non-Darwinian evolutionary change requires a course of development that is projectable such that given similar initial conditions in a similar environment it is rational to project similar results. It is unclear what short of laws provide an adequate basis for projectability, but this point need not be settled here. What matters is that there is no evident basis for projecting such patterns as do seem discernible in scientific change. On the contrary, science is remarkable for the variety and richness of its resources for solving problems, responding to challenges, and creating avenues for further growth. Any defensible generalization about how science develops will be too vague and qualified to serve as a basis for explanation or prediction.

This diagnosis passes over recent developments in cognitive science, including research in artificial intelligence and computational models of cognition.[4] If scientific discovery is interpreted as problem-solving and computers can be programmed to solve problems, then perhaps advances in science are governed by projectable rules.[5] If we abstract the right rules and successful science results from their further application, we establish projectability. It is questionable whether computers can be programmed to make scientific discoveries as opposed to reconstructing the reasoning by which scientific discoveries can be made. Discoveries are modeled by those who have already made them and who are thereby positioned to circumscribe and partition the space of possible solutions. But even if this computational approach succeeds in formulating general rules that guarantee results in a finite number of prespecifiable steps, there are two reasons why this success would not establish that science is evolutionary.

One reason is that the generation of such rules is itself a scientific advance that is not subsumable under its own results. It is clear that we will first have to know what the rules are if we are to write algorithms for their construction. The other is that the existence of rules for making progress in science is not evidence that science itself progresses by following rules. There is no reason to think that a rule-governed way in which science could change is anything like the way it does change. Rule-governed change is artificial.

Some models of scientific practice rich enough for explanation and prediction may not be subject to these objections. Bayesian models generate very specific results from seemingly modest and unproblematic

---

4. See Thagard 1988 and Churchland 1989.
5. See Newell and Simon 1972.

assumptions. There is considerable evidence, however, that scientists do not reason with probabilities in the way that Bayesianism requires. In particular their reasoning seems to be more causal than diagnostic; asked to reason in reverse, as required to apply Thomas Bayes's theorem, they go wrong, by Bayesian standards, systematically.[6] Also scientists do make decisions as to the acceptability of hypotheses and theories. All the Bayesian models let them do is assign probabilities.

Satisficing models make sense of definitive judgments of acceptability. They also accommodate better the ways complex social and psychological interests influence decision-making; Bayesian models relegate these factors to the subjectivity of prior probabilities. But satisficing models require unrealistic assignments of probability. Either there has to be a probability that the correct theory is among those under evaluation, or there must be a probability that experimental results will favor one contending theory over others. It is unclear where such probabilities come from, nor is there evidence that scientists think in such terms. Moreover, the scientific practitioner in a satisficing model is a highly idealized individual who is willing to accept the experimentally successful theory regardless of conflicts of interest or commitment.[7] The best one can say of the prospects for identifying projectable patterns in scientific decision-making, as a non-Darwinian evolutionary model requires, is that they are undecided.

4. REVOLUTIONARY MODELS

Revolution is a political concept. To revolt is to oppose prevailing authority by force or coercion rather than through processes sanctioned by established political institutions. A revolutionary model of social change has these features:

1. *Authority.* There is a dominant political authority demanding allegiance and controlling behavior in ways that preempt or dissipate challenges.

---

6. See Kahneman et al. 1982.
7. See Ron Giere's (1988) discussion of satisficing.

2. *Competing Interests.* There are groups whose interests conflict with the interests of the dominant authority.
3. *Force.* Opposition by force or coercion to the dominant authority shifts political power to a new authority.

To apply this model to scientific change, we identify the prevailing authority with a dominant theory or research program that dictates what problems are important and what lines of research are productive. This theory imposes standards for the evaluation of solutions, and these standards insulate the theory against internal opposition. In effect, the force and legitimacy of evidence are decided by the theory itself, so that evidence against the theory is either unrecognizable by or uncompelling to its practitioners. A theory dominates not only by the acceptance of its empirical commitments but also by establishing guidelines and methods for the conduct of research that protect its empirical commitments against the possibility of refutation.

Competing interests are represented by rival theories with different research agendas. Rivals to the dominant theory offer different guidelines, identify different problems as important, and measure research productivity differently. By the standards of a rival theory, the dominant theory is misdirected. What the dominant theory counts as influential achievements legitimizing its hegemony a rival dismisses or reinterprets. And symmetrically, by the standards of the dominant theory the rivals are nonstarters.

Such rivalry admits of no neutral or independent adjudication. Adjudication requires a standard of merit, and any such standard is embedded in and thus subservient to the interests of one of the contenders. Thus resolution requires force. By co-opting resources and gaining control of scientific institutions, those invested in a rival to the dominant theory manage to gain ascendance and shift the allegiance of the scientific community to their cause. There ensues a new period of domination in which not only is the content of science different but so are its standards and methods. Scientists divided by revolution cannot recognize one another as engaged in a common activity.

The key to representing scientific change as revolutionary is to regard it as essentially political. If scientific change is revolutionary, then the direction of science is determined by scientists' assessments of their own political interests. They calculate that allegiance to one theory or research program will better advance their careers than allegiance to another. They do not choose allegiances by determining that one theory is true and another false or that one is better supported by evidence than another. Those assessments would require a standard independent of any allegiance. The only basis for choice available independently of theoretical allegiance is the pragmatic calculation of personal interest. So scientists interpret evidence and invest credence in the way they judge their prudential interests best served. On the revolutionary model, what we take to be scientific knowledge are the empirical commitments of whatever theory currently dominates, unless we are in a period of undecided and shifting allegiances when either there is no knowledge or there are multiple, incompatible versions of knowledge.

How well does the revolutionary model describe scientific change? The original defense of this model proposed it as the alternative to a cumulativist picture, on which change adds to the existing body of knowledge without abandoning or disrupting what has already been achieved. Thomas Kuhn (1970) showed that this cumulativist picture trivializes scientific theories and lacks the resources to distinguish among past theories between those that were progressive and those that were misdirected. The traditional positivistic analysis of science interpreted scientific theories on the model of formal systems and assumed that successive theories could be compared as to their empirical success, much as logical or geometric theories are compared as to their power to generate theorems. Logical empiricism contended that the degree of evidential support afforded to a theory by its empirical success was fixed by a priori, analytically true inductive rules. This program is epitomized by Rudolf Carnap (1950). In opposition to Carnap, Kuhn argued that successive theories are not subject to any common formal measure of empirical support, a result he labeled "incommensurability." Kuhn contended that replaced theories are not merely improved upon but are refuted by their successors. However, the refutation depends on adopting the standards of the newer theory. A theory does not get refuted by the standards that measured its own success but only by the standards of an incommensurable theory.

Although Kuhn's criticisms of the positivistic tradition are surely correct and sufficient to discredit simple cumulativism, they do not enthrown revolution as the alternative. For we can agree that scientific change often consists in the refutation of theories without regarding refutation as a revolutionary process that changes the rules for reckoning success. Indeed it is difficult to see how simple refutation could be a revolutionary process. Refutation by the standards of a new theory is not refutation unless the new theory operates with the right standards. Refutation depends on some notion of what standards are right, some standard for standards. If it does not initiate a pernicious regress, any such notion will introduce standards of rationality and correctness independent of theory. Rejection and replacement could be revolutionary processes, but these can occur without refutation.

Kuhn himself contended that scientific change is often progressive just because rejected theories are refuted. It is unacceptable, he argued, to treat the inadequacy of Newtonian mechanics as one of limitations of applicability rather than as one of falsity. Newtonian mechanics is not a special case of relativity or reducible to it. The continuing predictive utility of Newtonian mechanics is not to be taken for correctness, even limited correctness. Although successfully applicable over a great range and variety of experience, Isaac Newton's laws are in fact false. They require the independence of mass from velocity. They preclude any limit on increase in velocity through the continuing application of external force. And they require instantaneous transmission of gravitational force.

But of course the unacceptability of these consequences presupposes the correctness of the relativistic ideas that replaced Newton. According to the revolutionary model, there is no standard independent of the contending perspectives that identifies one as correct. Or rather, any such standard would itself presuppose some further perspective whose correctness in turn is not independently establishable. So it is not clear how the emergence of a new position from the perspective of which a rejected theory has been falsified constitutes progress. In fact Kuhn has relativized the notion of progress in science to the standards of individual theories, just as refutation has been relativized. Progress by the standards of a theory is not progress unless the theory operates with the right standards. In denying the possibility of theory-independent standards, Kuhn denies us the conceptual resources to understand how science progresses.

Kuhn does have a theory of progress. Progress cannot be understood as advancement toward a goal, especially not an epistemic goal, because there can be no theory-independent measure of that. Instead, Kuhn suggests that science is progressive in the way that evolution is progressive; progress is to be judged not by advancing a goal but relative to previous stages. However, it is not clear that this way of assessing progress is open to one who regards changes of stage as revolutionary. For it is only by the standards of the later theory that it does improve over its predecessors. Whether improvement by these standards is really improvement, whether progress has really been made, depends again on whether these are the right standards.

Kuhn argues that revolutions are necessary to the development of science. As science is not cumulative, theories must be rejected. As theories are never rejected by their own standards, a new, incommensurable theory is required to reject them. But the choice among incommensurable theories is not rationally defensible without circularity. Accordingly, the explanation of theory change must appeal to social forces and, generally, interests external to science. Only these interests can force science to change, and scientific change, on Kuhn's view of it, must be forced. This argument is inconclusive, because it presupposes the theory dependence of all standards. If rejections are reasoned, they are not revolutionary. The failure of positivistic formal measures of the successfulness of theories does not show that theory choice is unamenable to reason. It does not show that theories are incommensurable in a way that makes the replacement of one by another revolutionary.

Kuhn's full position on the nature of theory change is actually more radical than that suggested by the argument I am criticizing as inconclusive. For Kuhn will not allow that theory *choice* is strictly possible. Choice requires an informed comparison of attributes, whereas, according to Kuhn (1970, 1977), the only attributes of rival theories understandable and recognizable by their advocates are empirical applications demonstrable in the laboratory. Like foreign languages, rival theories are conceptually inaccessible to their respective proponents. Understanding a theory requires adopting a particular way of thinking, and thinking in this way prevents one from thinking in the different way that understanding a rival theory requires. Thus mutual understanding is limited, communication is partial at best, and misunderstanding is the norm. A change of theoretical allegiance is therefore much more like a political or a religious conversion than like a decision or a choice.

Thus for Kuhn theory change requires conceptual revolution. And in fact scientific change often introduces revolutionary ideas that radically alter our understanding of the world. Changes in assumptions so deep that we did not even realize we were making them have often proved necessary for science to develop. The absolute character of temporal concepts is probably the most familiar example, because its relativistic replacements are clear. More radical is the subversion by quantum mechanics of our most basic ideas of the causal structure of the world, for we have yet to develop an understandable alternative.

Again, however, the phenomena do not sustain the revolutionary model used to interpret them. The admittedly revolutionary character of conceptual innovation does not make science fit that model, because revolutionary concepts can be adopted rationally. In adopting a new theory one thinks differently, but the doctrine that there cannot be reasons for doing so or that such reasons as there be must surpass understanding has little basis in scientific experience. It is supported not by history but by philosophy—by philosophical doctrines about meaning and language that embed scientific concepts in a nexus of theoretical commitments such that any change in theory changes the meanings of concepts needed to understand theory. Freed from these optional philosophical underpinnings, conceptual innovation makes science revolutionary only in a sense other than that which opposes the evolutionary picture.

5. COMPARISON OF MODELS

Evolution and revolution represent antithetical extremes. It is plausible that most forms of change will fit under neither model very well.

Change need not proceed continuously in accordance with general laws or rules, nor need it be so disruptive that the condition it replaces bears no clear connection to the condition it creates. Sometimes "evolution" is used as a synonym for "change," as though change as such, even revolutionary change, is automatically evolutionary. Even Kuhn repeatedly speaks of scientific revolutions as contributions to and mechanisms for the "evolution" of scientific institutions. But to the extent that institutions develop through revolution, they are *not* evolving. Revolution is far more common as a form of political change than scientific change. Though neither law nor projectable pattern governs change in science, the outcome of even radical scientific change bears clearly understandable relations to the conditions it supplants. Though controversial and indecisive during periods of change, the reasons for change are reconstructable and assessable independently of any commitment to its progressiveness.

In fact scientific change is cumulative, though not in the simple way that Kuhn rightly disputes. Successive theories may differ substantially in the concepts they employ, the problems they solve, and the questions they raise. We cannot measure scientific advance simply by the solution of problems or the addition of information to what was known before. Neither, however, does scientific change neglect outstanding problems or leave what were previously considered important facts about nature unexplained. If problems are to be dismissed rather than solved, if empirical phenomena are to be ignored rather than subsumed under new explanatory laws, then reasons must be given to make sense of these changes in status.

The revolutionary model creates the impression that the attention span of science is limited and shifting. Science is subject to abrupt and unpredictable changes in its concerns and interests. Instead, we find in a later stage of science the resources to explain departures from the problems and priorities of earlier stages. It is not incumbent upon later science to fulfill the research agenda of earlier science. But it is incumbent on later science to show cause why that agenda should be abandoned in favor of a new one. Scientific change is cumulative in the sense that what was right about accomplishments not retained is recoverable from later theories.

This idea of cumulativity is looser and more complex than the idea Kuhn rejected. Kuhn saw that successor theories could not be measured against their predecessors and the improvements weighed and counted. He did not acknowledge how often the disparity is produced by the relegation of former problems to the domains of other theories or even other sciences. Science is responsible for incorporating past successes under future theories, but this responsibility falls on the discipline as a whole, not on the particular theory that replaces a once-successful rival. Science is not as simply and clearly cumulative as the positivistic tradition supposed, but it is far too cumulative to fit the revolutionary model.

Consider some of Kuhn's favorite examples. One is the abandonment by Newtonian science of the medieval quest to discover some force or natural tendency to sustain the motion of a projectile after its release. Newtonian theory completely drops what was a major research problem of the Aristotelian tradition. But this neglect is not simply a change of interests or values. Newtonian science explains why the quest for a cause of projectile motion was misconceived. That quest presupposes the Aristotelian doctrine of natural place, which is inconsistent with Newton's laws and which could not, in any case, survive the change to a heliocentric universe. In addition to being empirically more successful than Aristotle's conception of natural motion, the Newtonian conception explains why the Aristotelian conception failed in application to the case of projectile motion.

Another example is the failure of Antoine Lavoisier's theory of combustion to explain the similar properties of metals. Phlogistic chemistry attributed the similarities of metals to their possession of phlogiston, which they released in burning. If burning is instead the process of oxidation, metals need have no common component to account for their similarities. However, phlogistic theory must make everything that burns a repository of phlogiston, including materials like carbon that are dissimilar to metals. It offers no structural account of how possession of phlogiston produces observable physical properties and in this respect is defective as an explanation. More importantly, Lavoisier's theory is simply the wrong place to look for an account of the properties metals share. This is a matter for electrical theory. That the theory

Laviosier's replaced addresses a problem that Lavoisier ignored does not challenge the cumulativity of *science*.

Of course an adequate electrical theory incorporating the notion of free charges was not available when oxidation was first understood. But we must remember that there are always outstanding problems. Some of these will have had solutions in earlier theories we now reject. That we must now regard them as unsolved need not count against the progressiveness of replacement theories. Instead, they may be properly relegated to domains for which adequate theories are yet to be developed.

Another example concerns change in the concept of mass from Newtonian to relativistic mechanics. Mass is a constant of proportionality in Newton's second law; in special relativity it depends on velocity. But it is an exaggeration to portray the difference as simply the adoption of a distinct concept as if each theory develops its own concepts and so cannot be understood from the perspective of the other. Relativity treats mass differently, because it regards mass as a form of energy. It is energy that is conserved, not mass as such. The achievements of Newtonian science involving the concept of mass are recoverable in relativistic science by treating mass as a conserved quantity, which to a good approximation it is. None of Kuhn's examples of radical conceptual change violate the cumulativist thesis as I have formulated it.

6. OTHER KINDS OF REVOLUTION

If scientific change appears more revolutionary than I take it to be, this may be because it often *is* revolutionary in ways other than that described by the political model. In addition to revolutionary conceptual innovation, there are several forms of revolution that science may be thought to exhibit.

The return or recurrence of a period of time or epoch is revolutionary, as is the return to or recurrence of abandoned ideas or concepts. These are revolutions in the cyclical sense that we use to describe orbital motion. Copernicus, who precipitated a revolution in astronomy by writing of the revolution of heavenly bodies, famously emphasized the ancient roots of his heliocentric model. Initially controversial because of the failure to find a model for the mechanism of gravitational attraction, gravity had by the mid-eighteenth century become accepted as an innate, irreducible property of matter. This conceptual shift represented a return to Scholastic explanations of gravitational motions in terms of inherent tendencies and essential properties of matter, which seventeenth-century corpuscularians had rejected. Relativity's subsequent analysis of gravity in terms of the geometry of space is strongly Cartesian in spirit. Old ideas often reappear in new guise, heralded as discoveries. But it would be as excessive to suggest that what happens in science is never truly new and different as it is to suggest that what happens bears no rational connection to the past.

7. CONCLUSION

Science appears revolutionary because the picture of the world that it gives us is ever more distant from common sense. Science introduces ways of conceptualizing natural processes that differ radically from expectations founded on ordinary experience. Additionally, there have been changes of theory so sweeping and so disruptive of social institutions invested in earlier theory that the term "revolution" does not seem excessive in describing them. Thus we speak of the "astronomical revolution," the "Darwinian revolution," the "chemical revolution," and the revolution in geology that replaced contractionist with mobilist models of the formation of the earth's crust. Calling these episodes "revolutions" is misleading, however, if it suggests that the political model of revolutionary change applies to them. The respects in which these changes are revolutionary do not preempt rational explanations of them that give us noncircular reasons to believe that they represent epistemic progress. They occur because there is reason to believe that older ideas are mistaken in fundamental ways, not because scientists reassess their political interests.

Science appears evolutionary because almost any change can loosely be so called and more importantly because science contains highly developed norms and standards strong enough to forge consensus over highly diverse ideas and interests. These standards regulate directions of scientific growth to an extent that makes science appear to develop autonomously, with its own internal mechanisms of change. This is an evolutionary picture in the broadest sense.

However, neither the revolutionary nor the evolutionary appearance of science sustains the epistemological implications of the respective revolutionary and evolutionary models of change. Science is too cumulative to be revolutionary and too unpredictably discontinuous and innovative to be evolutionary.

References

Carnap, R. 1950. *Logical Foundations of Probability*. Chicago: University of Chicago Press.
Churchland, P. 1989. *A Neurocomputational Perspective*. Cambridge, MA: MIT Press.
Darwin, C. 1964. *On the Origin of Species*. Cambridge, MA: Harvard University Press, 1964.
Giere, R. 1988. *Explaining Science*. Chicago: University of Chicago Press.
Kahneman, D., P. Slovic, and A. Tversky., eds. 1982. *Judgment under Uncertainty: Heuristics and Biases*. Cambridge: Cambridge University Press.
Kuhn, T. 1970. *The Structure of Scientific Revolutions*. Chicago: University of Chicago Press.
Kuhn, T. 1977. *The Essential Tension*. Chicago: University of Chicago Press.
Newell, A., and H. Simon. 1972. *Human Problem Solving*. Englewood Cliffs, NJ: Prentice-Hall.
Ruse, M. 1986. *Taking Darwin Seriously*. New York: Blackwell.
Thagard, P. 1988. *Computational Philosophy of Science*. Cambridge, MA: MIT Press.
van Fraassen, B. 1980. *The Scientific Image*. Oxford: Clarendon.

PART II

FOUNDATIONS OF PHYSICS

Chapter 7

What Can We Learn about the Ontology of Space and Time from the Theory of Relativity?

JOHN D. NORTON

1. INTRODUCTION

The advent of Albert Einstein's special and general theories of relativity in the first decades of the twentieth century changed philosophy of space and time. Prior to them, Euclid's ancient geometry and Isaac Newton's centuries-old notions of time and space provided a stable framework for philosophizing about space and time. Philosophers rarely challenged this framework. Rather, they asked, with Immanuel Kant, how we could reconcile the certainty of Euclid's and Newton's theories with the fragility of human learning. Or they asked, with Ernst Mach, whether the observational regularities summarized by these theories really licensed belief in the existence of entities—space and time—that elude direct experience.[1] It was disorienting when Einstein's discoveries showed that the absolute truths of Euclid and Newton were mistaken. What energized philosophical analysis into the deeper import of Einstein's

---

1. Did not Mach propose the new physical principle that inertia is caused by an interaction between bodies, and was it not used by Einstein to build his general theory of relativity? It is far from clear that Mach proposed that principle (see Norton 1995b) and that the principle is central to general relativity (see section 5).

theories was that he explicitly based his theorizing on philosophical reflections. New viewpoints soon multiplied, all supposedly vindicated by Einstein's theories, be they new findings on methods of discovery, on the nature of scientific theories, on the essence of space and time, on matter and cause, on being and bunkum.

My purpose in this chapter is to bring some order to the resulting surfeit. I will seek to answer the question of the title: what can we learn about the ontology of space and time from the theory of relativity? My task is not to survey the many answers on record. Rather, it is to find principled grounds for sifting among them and separating out a consistent, supportable view. What will aid in this task are some simple patterns in the ways Einstein's theories have been misinterpreted. For example, much of the philosophical analysis of Einstein's theories exaggerates the differences between relativity theory and the classical theory of space and time it replaces. Philosophers too often claim to find morals in relativity theory that could equally have been drawn from earlier theories.

Four Requirements

This concern motivates the first of four requirements that I shall ask all ontological morals to meet:

1. *Novelty.* The morals we draw should be novel consequences of relativity theory. They should not be results that could have been drawn equally from earlier theories.

The advent of relativity theory has allowed us to discern possibilities we just overlooked in the past. These are not morals of relativity. To say they are confuses message and messenger.[2]

2. *Modesty.* The morals we draw should be consequences of relativity theory. They should not be results we wish could be drawn from relativity theory but are only suggested to us by the theory.

---

2. To get a sense of how much interesting philosophy of space and time could be done prior to relativity theory (and after), see Sklar 1976.

Relativity has inspired many programs of research into space and time that are based on ontological themes. The hope is that our next great success in the physics of space and time will verify them. Unless these themes are consequences of relativity theory, however, they are not morals to be drawn here.

3. *Realism.* Relativity theory is to be construed as literally as possible.

We cannot draw ontological morals from relativity theory at all unless we take a particular attitude to the theory. In so far as is possible, we must take the theory to mean literally what it says; this is my favored formulation of realism. We are not compelled to adopt realism. But without it, there is no rhyme or reason in answers to the question of the title. We could choose to be fictionalists. Then we would judge the ontological pronouncements of relativity theory, whatever they might be, as useful mythmaking, devoid of insight into that which exists.

4. *Robustness.* We should not draw morals in one part of the theory that are contradicted in others. In particular the morals we draw from examination of special relativity should survive the transition to general relativity.

Failure to heed "robustness" has caused much unnecessary confusion. Many of the philosophical responses to relativity theory look at the special theory alone and trumpet results that are almost immediately contradicted by the emergence of general relativity.

1.2. Things to Come

The morals that I will discern all pertain to the theme of entanglement.[3] They will be elaborated in the sections to follow. In section 2 I describe how special relativity brought a new entanglement of space and time, and I will go to some pains to formulate the entanglement in a way that

---

3. I do not intend "entanglement" in the technical sense in which is has come to be used in philosophy of quantum mechanics. I use the term merely to designate the existence of rich and unexpected relationships.

respects *robustness*. In section 3 I review the most obvious moral that came with the extension of the special theory to the general theory, the entanglement of spacetime the container and the matter it contains. In section 4 I turn to the moral that has only been pursued seriously in more recent years, the entanglement of spacetime and causality, which forces any serious philosophical analysis of causation to examine what Einstein wrought. Finally, in section 5 I examine some of the popular claims that have not entered my select compendium and explain why I have spurned them.

2. THE ENTANGLEMENT OF SPACE AND TIME

2.1. The Relativity of Simultaneity...

As Einstein grappled with the problems in electrodynamics that gave us the special theory of relativity, the discovery of one misapprehension about space and time was key. It allowed him to reconcile two apparently incompatible notions, the principle of relativity, demanded by experiment, and the constancy of the speed of light, demanded by Maxwell's electrodynamics. He could assert both if he was willing to suppose something utterly at odds with classical theory: that observers in relative motion may disagree on which spatially separated events are simultaneous. Two events judged to occur at the same time by one observer might be judged to be sequential by another in motion with respect to the first observer. This result is the relativity of simultaneity. It is described in careful detail in the first section of Einstein's celebrated "On the Electrodynamics of Moving Bodies" (1952a), first published in 1905, for all that follows in Einstein's paper depends on it. It expresses a profound entanglement of space and time, a moral worthy of inclusion in our catalog.

This notion lay behind Hermann Minkowski's (1952, 75) immortal declaration when he introduced the concept of spacetime: "The views of space and time which I wish to lay before you have sprung from the soil of experimental physics, and therein lies their strength. They are radical. Henceforth space by itself, and time by itself, are doomed to fade away into mere shadows, and only a kind of union of the two will preserve an independent reality." Prior to relativity theory, space and time were treated separately. One considered space at one instant of time and then successive spaces as the instants passed. Minkowski combined these into a single four-dimensional spacetime manifold of events. That much was not incompatible with classical theory. In it, just as in relativity theory, the set of all events in space and time form a four-dimensional manifold. On pain of violation of *novelty* we cannot claim it as a moral of relativity theory. It was just a classical possibility that was not exploited. The novelty lies in the way the new spacetime can be decomposed into spaces that persist through time. It reflects a new entanglement of space and time. In classical theory the decomposition is unique. There is one way to do it. In relativity theory, as shown in figure 7.1, each inertially moving observer finds a different way to slice spacetime into spaces. There is a different decomposition associated with each inertial frame of reference.

Each such space consisted of *simultaneous* events. Since these observers could not agree on which events are simultaneous, they cannot agree on how to form the spaces. We cannot select one slicing as the correct slicing. Each is geometrically identical, and any criterion that would elevate one would do the same to all the rest. It is just like seeking diameters of a perfect circle. There is no one correct diameter that bisects the circle. There are infinitely many, and they are all identical in their geometric properties.

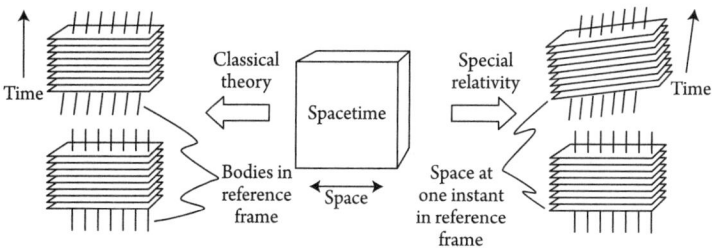

Figure 7.1. Simultaneity in classical theory and special relativity.

2.2 ... Is Not Robust

Important as it is, these traditional presentations of the relativity of simultaneity cannot stand as a moral, since they fail the requirement of *robustness*. With the advent of the general theory of relativity, spacetime took on a more varied geometric structure. It could still be sliced in many ways into spaces that persist with time, but in important cases just one slicing is preferred geometrically. To visit some familiar examples, consider the Robertson-Walker spacetimes used in standard big bang cosmology. Just one slicing gives spaces filled with a homogeneous matter distribution. Any other slicing mixes events from different epochs with differing densities of matter. Or consider a Schwarzschild spacetime, the idealized spacetime of our sun. There is just one natural[4] slicing, and it turns out to give us spaces whose geometric properties remain constant with time. Analogously, many chords might bisect the area of an ellipse, but bisection along the principle axis is geometrically distinct from all the others.

2.3. Infinitesimal Neighborhoods of Events in Classical and Relativity Theory

What of the celebrated entanglement of space and time brought by the special theory? Has the general theory parted what the special theory had joined together? It has not. To find a robust entanglement, we must seek it in a more subtle way. We will find it by exploiting a fundamental fact about the spacetimes of both the special and the general theories. They differ in domains of any finite extent. However, if we select just one event and consider the events infinitesimally close to it, then we have found a mini-spacetime that is the same in both special and general relativity. This mini-spacetime mimics the bigger spacetime of special relativity. The entanglement of space and time of the relativity of simultaneity can be found in it. We can use it to formulate this entanglement in a way that is robust under the transition from special to general relativity.

---

4. That is, natural in the sense that the slicing satisfies the technical condition of orthogonality with the world lines of the matter of the sun and the field's natural rest states.

ONTOLOGY OF SPACE AND TIME

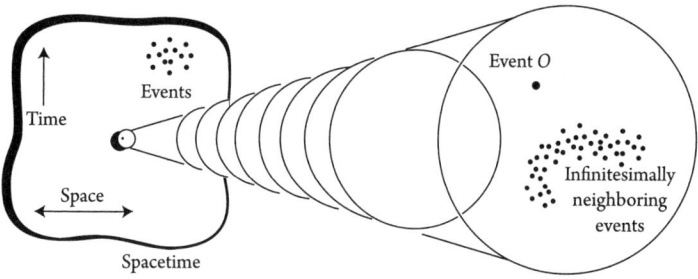

Figure 7.2. Extracting an infinitesimal neighborhood in both classical and relativistic spacetimes.

Let us proceed to these mini-spacetimes. In order not to violate *novelty*, I will first construct as much of them as I can in a way that is compatible with both classical and relativistic theories. In both the set of all events forms a four-dimensional manifold. That means that we can label events with four real numbers, the spacetime coordinates, and we can then use those numbers to decide which events are near which. This notion of nearness lets us extract the mini-spacetime of events infinitesimally neighboring some arbitrary event O, as shown in figure 7.2.

In the mini-spacetime we can identify events, such as T, that come temporally later than event O. A definite amount of time will elapse between events O and T as we pass along the spacetime trajectory OT. That time is physically measurable, for example, by counting the ticks of a clock that moves along the trajectory OT. Similarly, we can find an event S that occurs simultaneously with O (for at least one observer). The distance along the interval OS is physically measurable. For example, we might contrive a measuring rod to pass through events O and S so that opposite ends occupy events O and S simultaneously (at least for one observer). In the mini-spacetime the trajectory OT represents an inertially moving body and the interval OS a straight line.

We may also sum intervals represented by OT and OS using the familiar parallelogram rule for vectors. If we add OT and OS, as shown in figure 7.3, we arrive at OT'. If OT is the trajectory of some body, then

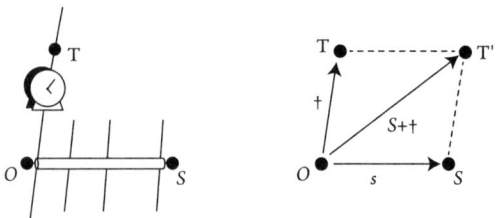

Figure 7.3. Measuring time elapsed and spatial distance between infinitesimally close events; the addition of displacements between infinitesimally close events.

OT' will represent the trajectory of a body moving in the direction of OS with respect to the first body.[5]

2.4. How Classical and Relativistic Spacetimes Differ: The Entanglement

Classical and relativistic spacetimes differ in the disposition of measurable times and distances in these minispaces. In the relativistic case,

---

5. A footnote for experts who suspect a sin against mathematical rigor in the "infinitesimal" talk: the mini-spacetime surrounding event O is really the tangent vector space at event O in the manifold. So displacements **t** = OT and **s** = OS are really tangent vectors. The times elapsed and spatial distance along the displacements (squared) are really the norms of the corresponding vectors, using the appropriate geometric structure. In the case of relativity theory, the norms are supplied by the metric tensor g. In the case of the Newtonian theory, I use a Cartan generally covariant formulation. The norm of **t** is derived from the absolute time one form dT; the norm of **s** is derived from the degenerate spatial metric h. Talk of a mini-spacetime of infinitesimally neighboring events is not so misleading, however. The tangent space at O can be mapped onto a neighborhood of O in the manifold of events by such maps as the exponential map. The intervals OT and OS represent inertial motion and spatial straights, because the exponential map assigns the vectors **t** and **s** to events along geodesics through the event O. By mapping onto neighborhoods of arbitrarily small size, one can come arbitrarily close to the geometric properties claimed for them. The mini-spacetime mimics the full Minkowski spacetime of special relativity in so far as this mapping need not be restricted to arbitrarily small neighborhoods to recover the properties claimed. The map can be from the tangent space to the entire Minkowski spacetime. That is, select a Lorentz normal coordinate system with origin at O in which the metric is g = diag (1, -1, -1, -1). Vectors **t** = (t, 0, 0, 0) and **s** = (0,

ONTOLOGY OF SPACE AND TIME

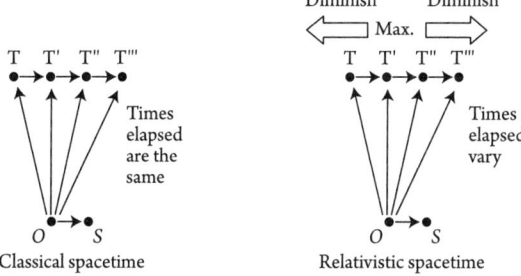

Figure 7.4. Entanglement of time and space in relativistic spacetimes.

times measured are entangled with distances in a way that they are not in the classical case. The difference is illustrated in figure 7.4. We take an arbitrary displacement OT between successive events. We add successively spatial displacements OS to it to recover new displacements OT', OT'', OT''', and so forth. In the classical case, the time elapsed along the displacements OT, OT', and so forth will all be the same. In the relativistic case they will differ. One will be the greatest value (labeled "max." in the figure), and the times elapsed will diminish as we proceed in either direction. (This decrease can continue until the time elapsed has dropped to zero, in which case the displacement represents the trajectory of a light pulse.)

This relativistic effect in the appropriate circumstance is equivalent to the time dilation (slowing) of clocks in relative motion. The clock at rest follows trajectory OT'. The remaining clocks move with respect to it along trajectories OT, OT'', and so forth and record a shorter time elapsed.[6] This

---

s, 0, 0) are mapped to events T = (t, 0, 0, 0) and S = (0, s, 0, 0) in the manifold, so that the metrical time elapsed along the geodesic OT is t and the metrical distance along the geodesic OS is s. The addition of vectors **s** + **t** corresponds to translation from O to event S and then translation by (t, 0, 0, 0) to arrive at event T' = (t, s, 0, 0). The distance OT' corresponds to the $\sqrt{\text{norm}}$ of **s** + **t** and is $\sqrt{t^2 - s^2}$. This last correspondence exploits the flatness of a Minkowski spacetime and in general precludes the vector space mimicking arbitrary spacetimes in general relativity.

6. Another footnote for the experts: The result is really that the Newtonian theory uses separate structures, dT and h, to determine times elapsed and distances, where relativity theory uses a Lorentz signature metric g for both. So taking the vectors **t** and **s** above, the Newtonian structures will assign the same norm to **t** and **s** + **t** since dT(**t**) = dT(**s** + **T**). In

entanglement is also closely related to the relativity of simultaneity; it can be used to generate a form of the relativity of simultaneity restricted to the mini-spacetime. For the details, see the appendix.

2.5. Other Candidates

The analysis above captures as best I can the sense in which space and time are entangled in relativity theory and in a way that respects the requirements of the introduction. The literature is thick with other proposals that are intended in greater or lesser extent to capture this particular novelty of relativity theory. I explain why I find some of the more prominent wanting.

2.6. Time Is the Fourth Dimension

This notion fails the requirement of *novelty*. Events, points in space at a particular time, form a four-dimensional manifold. That just means that the set of events can be coordinatized by four numbers. This is true in both classical and relativistic theories. In both we can say that the transition from the three-dimensional manifold of spatial locations to the four-dimensional manifold of events requires an extra dimension associated with time. The assertion is banal.

One might try to rescue the notion from banality by urging that there is something more inherently four-dimensional about relativity theory. That is true. It arises from the entanglement of space and time in relativity. As we have seen, however, that entanglement involves something more than the four dimensionality of the manifold of events. It involves the measurable times and distances between events and how they become interrelated. The observation that time is the fourth dimension of this spacetime hardly captures this entanglement. Is the tacit claim that time is not just a dimension of spacetime but one that is

---

relativistic spacetimes the metric g assigns different norms to them, since $g(\mathbf{t},\mathbf{t}) \neq g(\mathbf{t+s},\mathbf{t+s})$. The entanglement lies in the metrical structure; the addition of a spacelike vector to a timelike vector alters the norm of the vector.

just like the three spatial dimensions? That is a falsehood. The temporal aspects of spacetime always remain distinct from its spatial aspects.[7]

Indeed "time is the fourth dimension" is a mischievous slogan. It inevitably misleads novices seeking to distill the essence of relativity. Its banal meaning is so obvious that they are drawn to seek a profundity in its connotations. In 1903 the Wright brothers set us free from the two-dimensional surface of our earth and allowed us to soar freely in the third dimension. That dimension had always been before us, but we could not exploit it. It controlled us until the Wrights liberated us. Did Einstein in 1905 and Minkowski in 1907 repeat the feat? Did we learn through relativity theory how to free ourselves from the shackles of a three-dimensional world and roam freely in a fourth dimension, time, that had always stood before us? Of course not.

2.7. *The Determinateness of the Future*

When Minkowski (1908) introduced the routine use of spacetime into physics, it seemed that this represented the victory of a particular view of time. Minkowski's spacetime represented all there was, past, present, and future, and all at once. Did this finally vindicate an idea whose pedigree traces back to Parmenides in antiquity: time and change are mere illusions? To draw this as a moral of relativity theory, however, violates *novelty*. The four dimensionality of the manifold of events is shared with classical theories.

Might there be something special in the nature of a relativistic spacetime that supports the illusory character of change? An ingenious line of analysis suggests there might be. The argument exploits the spacetime diagram in special relativity shown in figure 7.5. Inertial observer A will judge events A_1 and B_1 to be simultaneous. Inertial observer B moves with respect to A, and observer B judges a different set of events to be simultaneous with event B_1. It includes event A_2 in observer A's future.

---

7. Timelike and spacelike vectors remain distinct. We can of course use an imaginary time coordinate in special relativity—$x_4 = ict$—so that the line element becomes $-ds^2 = dx_1^2 + dx_2^2 + dx_3^2 + dx_4^2$. The symmetry of the four coordinates is an illusion. The first three coordinates are reals; the fourth is imaginary.

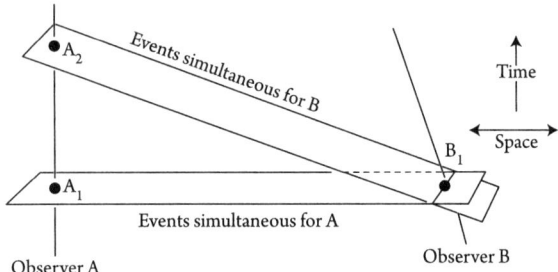

Figure 7.5. Determinateness of the future?

The argument is, then, that these judgments of simultaneity allow us to infer that event B_1 is determinate with respect to event A_1 and that event A_2 is determinate with respect to event B_1, so that overall event A_2 is determinate with respect to event A_1. That is, the future event A_2 is determinate with respect to the past A_1.

For our purposes the immediate problem is with *robustness*. The argument exploits the relativity of simultaneity, and that, we have seen, holds only infinitesimally once we pass to general relativity. Further steps are needed in the argument to establish the determinateness of events at large finite times in the future in the spacetimes of general relativity. If that problem could be remedied, we would face further difficulties. We are to accept that a judgment of simultaneity is sufficient for determinateness. We are to accept that the relation of determinateness is transitive when we are combining judgments of determinateness from different observers, even though judgments of simultaneity are not transitive in this way. Accepting both amounts to introducing new physical assumptions about determinateness into relativity theory. Thus the candidate moral violates *modesty* as well.[8]

2.8. Conventionality of Simultaneity

Einstein used the definition of figure 7.13 (see the appendix) to determine which events are simultaneous. Hans Reichbenbach interpreted

---

8. For an entry to the extensive literature on this question, see Maxwell 1993 and Stein 1991 and, for broader viewpoints, Čapek 1976 and Grünbaum 1976.

Einstein's use of a definition as revealing an important convention in the logical structure of relativity theory. If Einstein's definition really is just a decision on the use of a term, other uses could have been entertained. So, Reichenbach urged, any event between A_1 and A_3 at position A in space could be deemed simultaneous with event B_1 at distant position B in space; the choice is a matter of convention.

Might this conventionality be the appropriate expression for the entanglement of space and time in relativity theory? The proposal violates *robustness* in exactly the same way as the relativity of simultaneity, since the analyses of the conventionality of simultaneity are conducted in special relativity. A version of the conventionality thesis can be created in general relativity by mimicking the analysis in the minispace. As with the relativity of simultaneity, the conventionality fails if we relate the mini-spacetimes to the larger spacetime in so far as the larger spacetime can host a single preferred relation of simultaneity.

Aside from this problem, the claimed convention has been debated vigorously without a clear decision in favor of either side. The debate has been wide-ranging. In my view, its failure to be resolved results from lack of agreement on just what it takes to be a simultaneity relation. What is its physical meaning? Is it synonymous with determinateness—whatever that might be? What are its necessary formal properties? Must it be a transitive relation? Without clear answers, the debate meanders. In one reading that does entail transitivity, defining a simultaneity relation is equivalent to defining a time coordinate in spacetime, where the time coordinate cannot assign equal times to events that can be causally related. That trivializes the convention as merely a part of our broader freedom to choose coordinate systems arbitrarily. It also makes simultaneity conventional in cases in which it manifestly is not, such as in the spacetimes of standard cosmology. Yet the convention does tap into something important and novel in the spacetime structure of relativity theory: there are many more pairs of events that cannot be causally connected than there are in classical theory (see section 4). So should we say events are simultaneous just if they are not causally connectible? That violates transitivity and is less useful as a moral because of the ambiguity in the notion of simultaneity. Why not just take the greater freedom in lack of causal connectibility as the moral directly? Its meaning is clearer. If we take the stronger position that simultaneity, whatever it may be, is an inherently causal notion and so must be definable in terms of causal notions, then Einstein's simultaneity relation turns out to be the only nontrivial, transitive relation so definable. But why should we demand that simultaneity is so definable? For further discussion, see Sklar 1985, chap. 3; Norton 1999b; Grünbaum 2001; Janis 2002.

3. THE ENTANGLEMENT OF SPACETIME AND MATTER

3.1. Spacetime Loses Its Absoluteness

The general theory of relativity extends the special theory by the incorporation of gravitation. The standard approach had been to treat the gravitation field as a structure contained in spacetime, so that spacetime, the container, and the gravitational field, the contained, remained distinct. Einstein blurred this division of container and contained. The gravitational field became a part of spacetime itself. In the standard approach we say the earth orbits the sun because the gravitational field of the sun deflects the earth from the natural, uniform, straight-line motion dictated by spacetime. In Einstein's theory we say that the presence of the sun disturbs the geometry of spacetime, and that affects what are the natural motions for free bodies. Those natural motions now direct a freely moving earth to orbit the sun.

In adopting this new role, the character of spacetime was altered fundamentally. Formerly spacetime provided an immutable arena in which the processes of the world unfolded. This Einstein (1956, 55) characterized as the absoluteness of spacetime a characteristic special relativity shared with the older classical theory of Newton:

> Just as it was consistent from the Newtonian standpoint to make both the statements, *tempus est absolutum, spatium est absolutum*, so from the standpoint of the special theory of relativity we must say, *continuum spatii et temporis est absolutum*. In this latter statement *absolutum* means not only "physically real," but also "independent in its physical properties, having a physical effect, but not itself influenced by physical conditions."

In Einstein's new theory this absoluteness was lost. Spacetime is in turn altered by what it contains. The presence of the sun alters the geometry of spacetime in its vicinity.

In the remainder of this section I will review two manifestations of this entanglement of container and contained. The first is the failure of a particular view of the nature of spacetime. The second is the now less certain status of energy and momentum.

3.2. Spacetime Substantivalism

There is a natural division in the universes of general relativity. We have a four-dimensional manifold of events. And we have a metrical field defined on that manifold. Without that metrical field, we are unable to specify how much time or space elapses between events. The events of the manifold are like different colors in the rainbow. We can proceed smoothly though the colors: red, orange, yellow,... We can even see that orange is closer to red than yellow. But we cannot assign a distance in meters or a time elapsed in seconds to the passage from red to yellow. It is the same with the events of the manifold. The extra information of the metrical field tells us how much space or time lies between events as shown in figure 7.6. The aspects of the world that ordinarily we think of as gravitation are also encoded into this metrical information; its disturbance from the familiar Minkowskian disposition of special relativity is associated with the presence of a gravitational field.

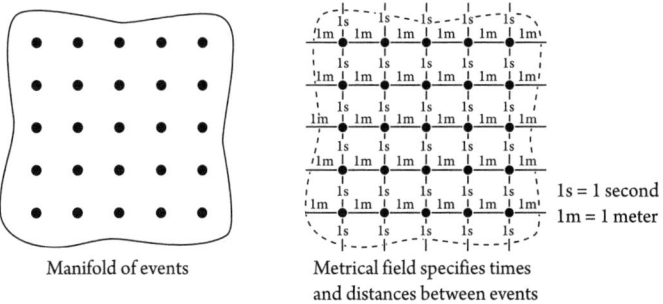

Figure 7.6. Manifold and metric.

Realism enjoins us to take this division seriously. The division should reflect some objective aspect of reality. The natural reading that does this is a version of spacetime substantivalism, manifold substantivalism. It identifies the manifold of events as spacetime, the container of the metrical field. Moreover, it attributes substance properties to the manifold; it has an existence independent of the fields it contains.

3.3. The Hole Argument

The hole argument first appeared in Einstein's work on general relativity toward the end of 1913. John Stachel (1989) recognized its nontrivial importance, bringing it once again to public attention. In its modern form it is used to ends different than Einstein's (see Earman and Norton 1987; Norton 1999a, 1999b). It shows that manifold substantivalism leads to some quite unpalatable conclusions that are usually deemed sufficient to warrant its dismissal. The hole argument exploits a property of general relativity, its general covariance. That property allows us to spread the metrical field across the spacetime in different ways. We may take the field and smoothly redistribute the same metrical properties over different events. We may effect this redistribution so that it occurs only in some arbitrarily designated region of spacetime—the "hole." Figure 7.7 shows the original and transformed metrical fields with the hole represented as a large circle.

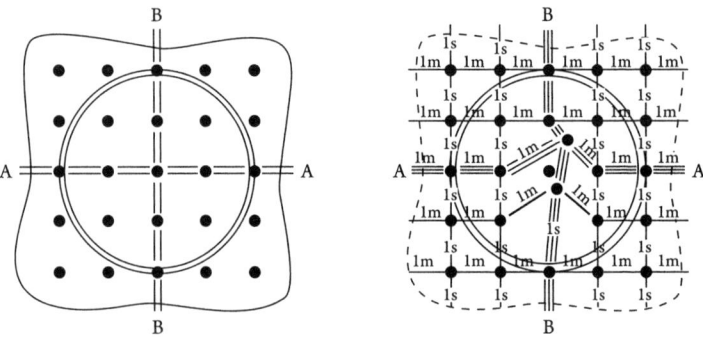

Figure 7.7. Transformation of the hole argument.

The two spacetimes are mathematically distinct. For example, imagine the straight line connecting AA picked out by the condition that it have the shortest distance and the straight line connection BB defined by the analogous condition that it have extremal elapsed time. The two straights will meet inside the hole. But they will meet at different events in the two spacetimes.

How are manifold substantivalists to interpret this difference? They are committed to the notion that the manifold of events has an existence independent of the fields defined on them; the events have their identities no matter what metrical properties we may assign to them. So the difference between the two spacetimes is a physically real difference for them. But it is a difference of a most peculiar type. It turns out that nothing observable distinguishes the two spacetimes. The times elapsed and distances passed to the meeting of AA and BB will be the same in both cases. Worse, everything outside the hole in the two spacetimes is identical; all the differences arise within. This is a failure of determinism of a most serious kind—the hole can be specified to be as small as we like. No specification of spacetime outside the hole can succeed in fixing its properties within. That is, the manifold substantivalist is committed to factual differences between the two spacetimes that are opaque both to the observation and to the determining power of the theory.

3.4. Entanglement of the Manifold of Events and the Metrical Field

The natural response is simply to assert that the differences between the two spacetimes are merely differences in mathematical description; they both describe the same physical reality. This widely accepted escape amounts to the rejection of manifold substantivalism. In particular we say that the meeting points of AA and BB in each case represent the same physical event, even though they are mathematically distinct point events in the manifold.

In rejecting spacetime manifold substantivalism, we see the entanglement of spacetime and its contents. Consider a universe with gravitation but no other contents. We cannot split off the manifold of events as the spacetime container from the gravitational field held within it in the metrical field. Which physically real events are identified by which mathematical point events of the manifold cannot be decided without consulting the information of the metrical field. As we spread that field differently over the mathematical point events of the manifold, we alter their physical identities.

3.5. Is It Novel?

The hole argument entered the literature as a result of Einstein's work on the general theory of relativity, and in its modern guise we infer from it that manifold substantivalism is untenable. But does it satisfy the requirement of novelty? On this there are differing schools of thought. They divide according to how one understands the requirement of general covariance that allows the transformation of the metric shown in figure 7.7. One view holds this to be a special feature of general relativity only. In special relativity, for example, the corresponding metrical structure is given once globally and is not subject to transformation (see Stachel 1993). In this view, the hole argument and its consequences satisfy the requirement of novelty. A second view, which I hold, allows that even classical theories may be formulated in a way that permits the transformation of figure 7.7. These are called "local spacetime theories" in Earman and Norton 1987. Under that view, the failure of manifold substantivalism is common to all theories, classical and relativistic, if they are appropriately formulated, so the failure does not meet the requirement of *novelty*.

3.6. The Problem of Gravitational Field Energy Momentum

In the classical view, space and time are the containers; matter is what is contained.[9] The distinctive property of matter is that it carries energy and momentum, quantities that are conserved over time. A unit of energy cannot just disappear; it transmutes from one form to another,

---

9. For a recent discussion of gravitational field energy momentum, see Hoefer 2000.

merely changing its outward manifestation. This property of conservation is what licenses the view that energy and momentum are fundamental ontologically. They are the stuff of matter. In the course of many interactions, they are the substances that persist, merely changing their form: the chemical energy of coal is transformed to the heat energy of the fire, to the pressure energy of the steam in the boiler, to the energy of electricity in the generator, and so on.

The reality of fields as a type of matter is in part revealed by their carrying of energy and momentum. The electromagnetic field, for example, carries the energy of the sun to us in the form of sunlight. That energy can be put to good use. Vegetation absorbs it and uses it to grow. We can use it to heat water in solar hot water heaters. Sunlight also carries momentum from the sun. As a result, when sunlight blazes down on us, its impact on us creates a pressure from the momentum imparted to us. As it turns out, that pressure is too small to be noticed by sunbathers. Otherwise the momentum carried by sunlight would be as familiar as the energy it carries that warms a chilled bather after a swim.

Classically, one would expect the gravitational field to carry energy and momentum as well. When great masses of water run down a mountainside and through a hydroelectric power station, electrical energy is produced. It comes from the energy stored in the classical gravitational field. The effect of the lowering of the water is to intensify slightly the earth's gravitational field. This intensified field has less energy. The lost energy is imparted to the falling water as kinetic energy and then to the generators in the power station. Similarly, the falling water carries momentum that was imparted from the gravitational field. If general relativity is to return a reasonable classical view of gravity in the case of weak gravitational fields, there must be some corresponding provision for gravitational field energy and momentum. The expectation is powerful, but general relativity has shown itself to be most reluctant to give gravitational energy and momentum the homage it draws classically.

3.7. Densities: Big T and Little t

In general relativity the density of *non*gravitational energy and momentum at an event in spacetime is represented by the stress-energy tensor of matter, represented symbolically by T—"big T." It is the structure that encodes the total energy and momentum densities due to all nongravitational forms of matter, such as fluids, solids, and the electromagnetic field. In his original work in general relativity, Einstein defined an analogous quantity, the stress-energy tensor for the gravitational field, represented symbolically by t—"little t." It was heuristically very important to Einstein, because he could use it to convince himself that both nongravitational and gravitational energy and momentum had the same power to generate a gravitational field. Indeed that the gravitational field's own energy and momentum generates a gravitational field is one way to see how the notorious nonlinearity of Einstein's theory arises. However, Einstein's little t caused a lot of trouble for the generations of relativists that came after him. The difficulty is stated most simply in mathematical language: big T is a true tensor, but little t is not; it is a pseudotensor. What this means is that is that big T can be represented independently of the particular coordinate system we use in spacetime, but little t cannot.

This might not seem to be such a problem until we realize that the choice of different coordinate systems is merely a choice of different ways to describe the same reality. What is real is what is common to all the descriptions. What varies from description to description is merely an artifact of the mode of description. If there is a nongravitational energy density at an event, big T is nonzero. No change of coordinate system can make big T vanish. It is different for little t. If we have a nonzero little t at some event, we can *always* choose a new coordinate system in which little t vanishes at that event (and conversely). It is as though we send out many different investigators to inform us of the energy density at some event. All will agree on whether there is a nongravitational energy density present; they will not agree on whether there is a gravitational energy density present. What are we to think of the density of gravitational energy and momentum at this event when we read these conflicting reports?[10]

10. Analogous problems arise in a formulation of Newtonian theory that represents gravitation as spacetime curvature (Wald 1984, 286n4). Does this mean that the effects described in the text violate *novelty*? I do not think they do. I take the association of gravitation with spacetime curvature to be the novelty of general relativity, and that novelty has been borrowed by the formulation of Newtonian theory at issue.

3.8. The Gravitational Energy and Momentum of Extended Systems

The standard response to this problem in the literature is that "the energy of the gravitational field cannot be localized" (see Misner et al. 1973, secs. 20.3, 20.4). We can only talk of gravitational energy and momentum of an extended system and not the density of gravitational energy and momentum at a particular event. In so far as I can understand this response, it really just tells us that little t should be given no physical interpretation. It should merely be used as a mathematical intermediary in computing the gravitational energy and momentum of extended systems.

The difficulty with this response is that the gravitational energy and momentum of extended systems fare only marginally better. (For elaboration of what follows, see Wald 1984, 285–295.) Following the classical model, one would expect that we could take the energy and momentum densities of big T and little t and sum them up over the space occupied by, for example, a galaxy of stars to find the galaxy's total energy and momentum. In general this cannot be done. One cannot define meaningfully the total energy and momentum of some extended system where the total energy and momentum is to include both gravitational and nongravitational contributions. At best these total quantities can be defined in special cases.

The summation of the information in big T and little t to recover a total energy can be done if there is a rest frame in which the geometry of the spacetime is independent of time. That would arise if we had a completely isolated galaxy not of stars but of passive lumps of matter held apart by sticks such that the whole system just sat there completely motionless.[11] Real systems are not so nicely behaved. Stars radiate and

---

11. Technically, the spacetime must admit a timelike Killing field ξ^a, which satisfies $\nabla_a \xi_b + \nabla_b \xi_a = 0$ and is tangent to the world lines of the frame mentioned. Then the differential law $\nabla_a T^{ab} = 0$ can be integrated to give a conserved quantity corresponding to the total energy. If there is a Killing field, then the differential law entails that $\nabla^a(T_{ab}\xi^b) = \nabla^a(T_{ab})\,\xi^b + T_{ab}\nabla^a \xi^b = 0$. This quantity $\nabla^a(T_{ab}\xi^b)$ can be integrated over suitable spacelike hypersurfaces Σ via Stokes's theorem. Following the usual procedure in which a boundary term is contrived to vanish, the energy of the system is recovered as $\int_\Sigma T_{ab}\xi^b n^a$, where n^a is a unit normal vector to the surface Σ. This energy is a constant

thereby change their masses; stars in galaxies move about relative to each other; gravitational waves impinge upon the galaxy from the outside. All this affects the geometry of spacetime and precludes the summation.

There is another circumstance in which the total energy of a gravitational system such as a galaxy can be defined, even if there is considerable internal change in the galaxy. That arises when we presume that the galaxy sits in a spacetime that becomes asymptotically flat as we travel to spatial infinity. That is, if we are far enough away from the galaxy in all directions of space, we find ourselves in spacetime that comes arbitrarily close to the Minkowski spacetime of special relativity. In such a spacetime we are able to define the energy of a system. We can use those abilities to define the energy of a distant galaxy, since we can treat that distant galaxy in largely the same way as we would a distant object in special relativity.

3.9. Energy, Momentum, and Force

In retrospect we should not have been taken aback by the compromising of energy and momentum in general relativity. The first thing that one learns in approaching general relativity is that the notion of force has been compromised. General relativity no longer offers a precise notion of gravitational force. It has been "geometrized away." But one cannot geometrize away force without further ramifications. Consider how intimately energy, momentum, and force are related. In classical theory if we have a constant FORCE acting on a mass for some TIME during which the mass moves through some SPACE, the ENERGY and MOMENTUM gained by the mass is related to FORCE according to:

$$\text{ENERGY} = \text{FORCE} \times \text{SPACE}$$

$$\text{MOMENTUM} = \text{FORCE} \times \text{TIME}$$

in the frame because of the vanishing of $\nabla^a(T_{ab}\xi^b)$. If there are corresponding spacelike Killing vector fields, then the total momentum of the system in the direction of the Killing field can be defined analogously.

If gravitational force has somehow been compromised—geometrized away—then we should expect the same to happen to the other dynamical quantities in these two equations, ENERGY and MOMENTUM.

Correspondingly, in those cases in which the classical notion of force is restored, we can define the energy of the gravitational system. The restoration of a Minkowski spacetime in asymptotically flat regions of spacetime allows us to use the resources of special relativity to reintroduce a notion of gravitational force. It is identified with the geometric perturbations of the metrical structure from the exact flatness demanded by a Minkowski spacetime.

3.10. The Infection Cannot Be Restricted to Gravitation Alone

Perhaps our hope is that we might localize the difficulty to gravitational dynamical quantities only. Gravitational force has been geometrized by general relativity, but electrodynamical force has not. So perhaps gravitational energy and momentum are compromised but not electrodynamical energy and momentum. This hope is quickly dashed once we recall the interconvertibility of all forms of energy and momentum. What classically gives energy and momentum their special status ontologically is that they are conserved through all change, as we saw above. Once we have the transformation of a nongravitational form of energy into a gravitational form, we are no longer able to assert the conservation of the energy and momentum in all generality. The total energy and momentum of the system is no longer well defined. These conversions are pervasive. We convert energy and momentum between gravitational and nongravitational forms every time a body rises and falls above the surface of the earth, as the earth proceeds in its orbit from aphelion to perihelion to aphelion, as a star undergoes gravitational collapse and releases its gravitational energy as radiation, and so on.

When these conversions proceed, the matter contained by space and time, as captured by its energy and momentum, loses its unequivocal character as content of space and time. It becomes entangled with its container in the structure of space and time itself.

3.11. All Is Geometry?

What moral do we not draw? Once we see how gravitation is geometrized in general relativity, it is easy to expect that all remaining forces will be geometrized as well. Einstein was the most prominent advocate of this view; his legendary quest for a unified field theory amounted to the quest for a theory that geometrized electrodynamical force just as general relativity had geometrized gravitational force. The ontological moral would be that everything is fundamentally a geometric property of space and time. While one cannot exclude future research establishing such a result, we do not have it now. To conclude this moral would violate the requirement of *modesty*.

4. ENTANGLEMENT OF SPACETIME AND CAUSATION

4.1. The Speed of Light

The most prominent fact about special relativity is the constancy of the speed of light. This constancy is somehow built into the essence of space and time according to the theory; spaces and times repeatedly contort themselves to preserve its constancy for all inertial frames of reference. We must of course recover from the novice error that there is something special about light that brings all this about. In principle light—electromagnetic radiation—has nothing to do with it. Special relativity would say the same things about space and time in a completely dark universe. The real result is that there is a special speed built into the structure of space and time, and in seeking to go as fast as it can, light happens to travel at that speed.

Are we to recover an ontological moral from the *constancy* of the speed of light? That might be possible were it not for general relativity. Any such moral would violate *robustness*. In general relativity there is no comparable sense of the constancy of the speed of light. The constancy of the speed of light is a consequence of the perfect homogeneity of spacetime presumed in special relativity. There is a special velocity at each event; homogeneity forces it to be the same velocity everywhere.

We lose that homogeneity in the transition to general relativity, and with it we lose the constancy of the speed of light. Such was Einstein's conclusion at the earliest moments of his preparation for general relativity. Already in 1907, a mere two years after the completion of the special theory, he had concluded that the speed of light is variable in the presence of a gravitational field; indeed, he concluded, the variable speed of light can be used as a gravitational potential.[12]

4.2. Causal Structure

While the constancy of the speed of light is not preserved in the transition to general relativity, the existence of a special speed at every event is preserved. To see it we need to return to the mini-spacetimes surrounding each event that were introduced in figures 7.2. In particular, if we continue to add spatial displacements to the event T, we will eventually arrive at a pair of successive events OL such that the time along OL will be zero, as shown in figure 7.8. There is a corresponding displacement OL' of zero time elapsed in the opposite direction.

The displacements OT, OT',... represent timelike displacements and are instantiated by things moving slower than light. The displacement

---

12. See Einstein 1907, 1952b. It is not so easy ask if the speed of light is constant in general relativity. At first it looks like the result survives. The metrical norm of any lightlike vector is zero, so that if this zero norm measures the speed, then it is always the same—although zero is an unusual measure for the greatest achievable speed. Also, in the neighborhood of any event, one can always set up measuring rods and clocks in free fall and of sufficient smallness, so that they measure the same constant for the speed of a light signal. However, there seems no general way to extend this constancy to measurements conducted over extended regions, as Einstein realized in 1907. Consider the simplest case of a static spacetime that can be foliated into a family of spacelike hypersurfaces with a time-independent geometry. We can use any physical process to assign times to the surfaces, a kind of cosmic clock. But in the general case there is no way to do this so that all light signals propagate through the spaces with the same speed on this cosmic clock. For details of these constructions, see Norton 1985, sec. 3. Einstein's view (as elaborated in Einstein and Fokker 1914, sec. 2) seems to have been that the constancy of the speed of light entails that the spacetime is conformally flat, such as it was in his reformulation of the Nordström theory of gravitation. The spaces of general relativity are not in general conformally flat, so that sense of the constancy of the speed of light fails.

FOUNDATIONS OF PHYSICS

OL represents lightlike displacements that are instantiated by things moving at the speed of light. The displacements like OS are spacelike; they connect events that some observers will judge to be simultaneous. This three-way division of displacements can be effected in the mini-spacetime of every event. If we add in an extra dimension of space, we find that the events lightlike related to the event O lie on a cone as shown in figure 7.8. The top half of the cone represents the events in the mini-spacetime surrounding event O passed by an expanding shell of light that originates in O. The bottom half represents the time reverse, the events passed by a spherical shell of light that collapses on O.

So far I have said nothing about whether a body can move so that is passes along a spacelike displacement OS. Relativity theory does not preclude such bodies; they would represent tachyons, particles that travel faster than light. It has long been standard to assume that such propagations are not possible. This is an additional but natural supposition fundamental to all that follows. What makes it natural is that any spacelike propagation produces temporal anomalies. Even if one observer judges a tachyon to be traveling just slightly faster than light, there will be another observer who will judge its propagation to consist of simultaneous events, so that it moves infinitely fast, and another that will judge its events to progress into the past, so that it moves backward in time. There is no incoherence in these judgments as long as we contrive our theory to prevent propagation into the past generating

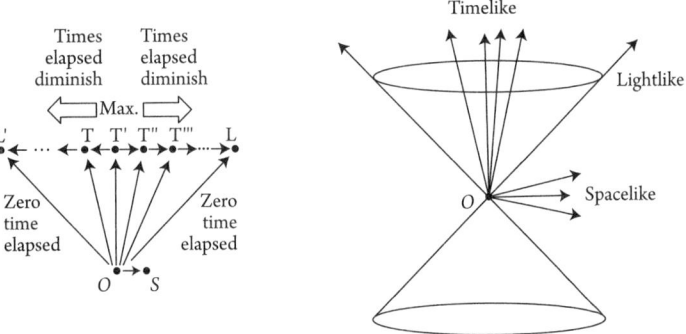

Figure 7.8. Light cone structure.

self-contradictory closed causal loops. We are protected from their realization by presuming that nothing moves along spacelike displacements. We are assured that this is not an unreasonable presumption, since we have no empirical evidence of tachyons and we have built the highly successful modern quantum field theory on the presumption that there cannot be spacelike propagations in special relativity.

4.3. The Relation to Causality as Normally Understood

It is standard in the physics literature to talk of the light cone structure as the causal structure of the spacetime. That designation can be misleading. General relativity does not have a fully developed metaphysics of causation such as would be expected by a philosopher interested in the nature of causation. Rather, we should understand the causal structure of a spacetime in general relativity as laying out necessary conditions that must be satisfied by two events if they are to stand in some sort of causal relation. Just what that relation might be in all its detail can be filled in by your favorite account of causation.

The condition for the possibility of a causal relation between two events is illustrated in figure 7.9. Event O and T of the full spacetime are causally connectible if they can be connected by a curve that is

Figure 7.9. Causal connectibility.

everywhere timelike or lightlike. (This means that at the mini-spacetime of every event on the curve, the curve corresponds to timelike or lightlike displacements.) When O and T are so connectible, a body can travel between them without ever exceeding the speed of light. Events O and S are not causally connectible if any curve that connects them must at some event be spacelike; a body that tries to travel between them must somewhere exceed the speed of light.

4.4. *Causal Isolation*

While general relativity only places necessary conditions on causal connectibility, they prove to be quite powerful. They make possible a far richer repertoire of causal connections while at the same time thoroughly entangling causal connections with the structure of space and time. In one part of this new repertoire we have universes that have much less causal connectibility than we would otherwise expect.

The fully extended, matter-free Schwarzschild solution is a kind of black hole and one of the simplest spacetimes admitted by general relativity. From outside the black hole its spacetime just looks like that of our sun. If one were to fall in, one would end one's journey in a singularity as the curvature of spacetime grew without bound. On the other side of the black hole is a second world with a geometry that exactly clones that of the first spacetime. Inhabitants of that new world can also fall in the black hole and even meet those who fell in from the old world, before they come to their end in the singularity. There is a counterpart to that fatal singularity. It looks like a black hole in reverse, a singularity that can emit, a white hole. The white hole singularity can causally affect both worlds. These worlds naturally combine into a single universe. However, the causally intriguing aspect of this universe is that there is no possibility of direct causal connection between the two worlds. There is no way for inhabitants of one to voyage to the other or even to signal to the other. They are causally isolated.

Another example pertains to the "horizon problem" in cosmology. In standard big bang cosmologies we can trace back the motion of matter in the universe only finitely far into the past. As we do the matter of the universe gets compressed to arbitrarily high densities in the approach to

ONTOLOGY OF SPACE AND TIME

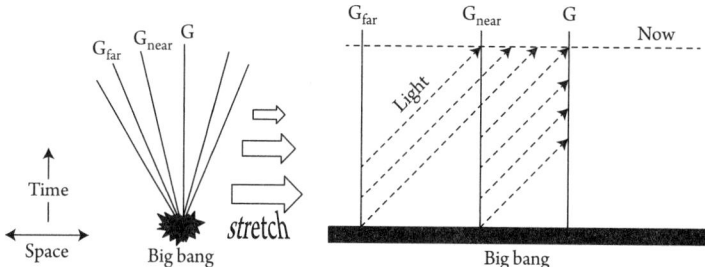

Figure 7.10. Conformal diagram of a big bang universe.

the big bang. Take the case in which we idealize the matter as particles—"dust." Because of this unbounded compression, one might expect our piece of dust at some time in the past to have had the possibility for causal contact with every other piece of dust. That turns out not to be so. If the expansion of the universe in the cosmology is sufficiently fast, a signal traveling to us at the speed of light from some nominated piece of dust can never arrive. We flee too fast. Our part of the universe is causally isolated from many others. This effect is most easily illustrated in what is known as a conformal diagram such as is shown in figure 7.10. The trajectories of the dust particles, really the galaxies, are shown on the left emanating from the big bang. That representation does not allow us to see what can causally connect with what. It is a simple trick to stretch out the diagram so that all lightlike propagation proceeds along lines at 45 degrees. The big bang now appears as a long stretched out band. We read from it that our galaxy G now could have been causally affected by galaxy G_{near} but not by galaxy G_{far}.[13] The boundary marked by the furthest galaxies that can affect G now is our "particle horizon."

---

13. For concreteness I have in mind a Robertson-Walker spacetime filled with pressureless dust at exactly the critical density so that its spatial sections are Euclidean. The invariant interval s is given by the line element $ds^2 = -dt^2 + R(t)d\sigma^2$, where $d\sigma^2$ is a Euclidean line element and the time coordinate $t > 0$. $R(t) = 0$ in the limit as $t \to 0$, which designates the big bang. In suitable units (Hawking and Ellis 1973, 138) for this most simple of cases, $R(t) = (3t)^{2/3}$. Introducing a new time coordinate $\tau = (3t)^{1/3}$, the line element becomes $ds^2 = R(\tau)(-d\tau^2 + d\sigma^2)$, so that the original spacetime is conformal to half a Minkowski spacetime $ds^2 = -d\tau^2 + d\sigma^2$, where $\tau > 0$.

This is just the simplest example of how horizons can separate off causally inaccessible parts of the universe. In other examples there remain portions of the universe that are causally inaccessible to us no matter how long we wait on our galaxy, even in the limit of infinite time.

4.5. Causal Abundance

The examples so far have shown us less causal connectibility in general relativity than we might expect. We can also have more and in ways that are traditionally of interest to philosophers.

In general relativity the trajectory of an observer through spacetime, the observer's world line, is dictated by the spacetime geometry. That geometry admits quite complicated structure and connections. In particular there prove to be many universes in which an observer's world line can be connected back to meet its own past. This can happen many ways. The simplest just involves a mathematical construction that is essentially identical to what we do when we roll a piece of paper into a cylinder by gluing its opposite edges. We can glue the future of the spacetime to its past and thereby produce a universe in which observers can meet their past selves merely by persisting long enough in time (see figure 7.11.

There are less contrived but more complicated ways of bringing about this possibility. In a Goedel universe cosmic matter rotates, and observers who accelerate sufficiently intersect their pasts. In other

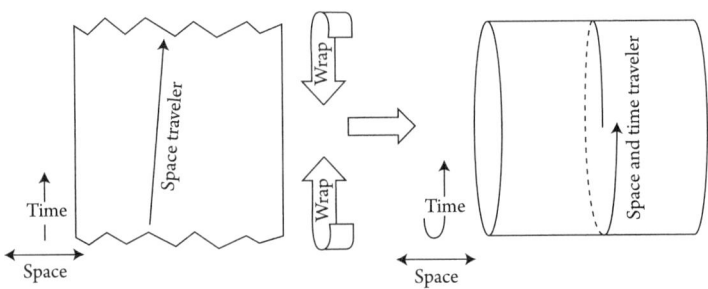

Figure 7.11. A universe with time travel.

universes we need only an infinitely long, dense rod of matter spinning rapidly to achieve the same end. Or in others we open up wormholes that connect one place and time with other places at different times; these are portals through which would-be time travelers can pass (see Earman 1995, chap. 6). If one understands "possible" to mean licensed by our best physical theories, then there can be no doubt that time travel is possible. That does not mean that there is time travel in our universe. Indeed a universe in which time travel actually occurs is likely to be much different from the one we are familiar with. It must be so contrived that present actions can only take place if they will cohere with the interfering machinations of a future time traveler with the past of those actions (see Arntzenius and Maudlin 2000).

In foundations of mathematics and computation, it is often taken as a commonplace that an infinity of discrete actions cannot be completed. Hence what is computable is restricted to what can be calculated in finitely many steps. If one understands "possible" to mean licensed by our best physical theories, then, at least as far as the spatiotemporal aspects are concerned, completing an infinity of computations is possible. In a sense to be explained, general relativity allows systems in which the completion of a quite ordinary infinity of manipulations is allowed. These arise in Malament-Hogarth spacetimes (see Earman and Norton 1993). The defining characteristic of such spacetimes is that they admit world lines for a slave-master pair. The slave persists for an infinity of time, perhaps fully occupied computing some uncomputable function; the master can be so located in the spacetime that, after finite time has elapsed along the master's world line, the master is able to see the entire infinite history of the slave's world line. If the slave is trying to determine if a given Turing machine halts on a given input by a simple simulation, the master will learn what the slave never learns assuredly at any finite time in the slave's life: whether the machine halts. For example, the slave may be set up to send a light signal to the master just in case the slave's program halts. At no stage of its infinite life will the slave know that the signal was sent, but the master will come to know this assuredly after a finite time of the master's. The slave and master are illustrated in a conformal diagram of a Malament-Hogarth spacetime in figure 7.12.

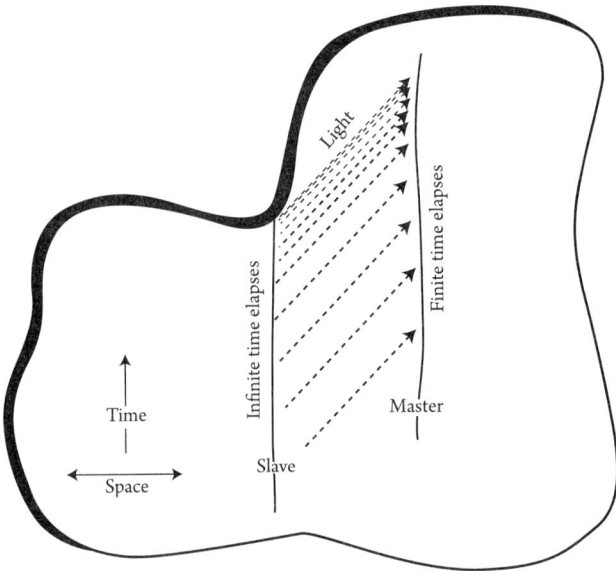

Figure 7.12. A Malament-Hogarth spacetime.

4.6. Causal Theory of Time

One of the best-known attempts to extract a fundamental causal moral from relativity theory is Reichenbach's (1956) "causal theory of time." Its central claim is that the spatiotemporal relations between events are reducible to causal relations: event A is earlier than B just means that event A could causally affect B. Might we find in this theory an ontological reduction of spatiotemporal structure to causal structure? I do not believe we can read this ontology of cause from relativity theory.

There are two problems. First, Reichenbach's analysis is dependent on a formal result. It is possible to axiomatize the special theory of relativity in terms of causal relations alone, so that the other spatio-temporal relations are derived relations. This formal result holds only in special relativity. It fails with the transition to general relativity. The failure is easy to see. Causal connectibility of events is just lightlike or timelike connectibility. In general relativity one can have many distinct spacetimes with the same relations of causal connectibility.[14] So an axiomatization in terms of causal connectibility alone cannot provide the extra structure needed to distinguish the two cases. As a result the causal theory of time violates *robustness*.

Second, that a spacetime theory can be axiomatized in terms of causal structure does not establish the ontological primacy of the entities taken as primitive in the axiomatization. To think otherwise creates great difficulties. There are many distinct axiomatizations possible for a given theory, and we cannot take all the primitives as ontologically primary on pain of trivialization. The decision of which axiomatization properly reflects the ontology is quite delicate. It seems natural to axiomatize Newtonian particle mechanics with the mass and velocity of the particles as the primitive notions and their energy and momentum as derived, although the reverse is also possible. This naturalness dissipates once one extends the particle mechanics in almost any way by, for example, adopting a Lagrangian or Hamiltonian formulation, or extending it to a field theory, or relativizing it, or quantizing it. Then energy and momentum appear more fundamental ontologically, with mass and velocity derived quantities. One might look to ontological significance in the simplest of axiomatizations. Such a principle is hard to implement without clear guides on how to assess simplicity. In any case, Reichenbach's and other causal axiomatizations contain large numbers of postulates and informally seem anything but simple. For further discussion, see Sklar 1985, chaps. 3, 9, 10.

5. RELATIVISTIC MORALS THAT FOUNDER

Einstein's theories of relativity are really theories of space, time, and gravitation, although that is not reflected in the names that Einstein gave his theories. He called them the special and general theories of relativity as a reflection of how he thought about the theories and how he came to discover them. The outcome has been a special emphasis on

---

14. Figure 7.10 illustrates such a case. The Robertson-Walker spacetime on the left has the same causal structure as a half Minkowski spacetime.

relativities of various sorts in attempts by later scholars to interpret the theories. The tendency has been for these relativities to be overemphasized, so that morals derived from them are often unsustainable as novel lessons of relativity theory. I review a few examples.

5.1. "All Is Relative"

Need I warn anyone with a modicum of philosophical sophistication that this weary slogan gains no support from relativity theory? The relativity Einstein found in his theories is a relativity of measured quantity to observer. So the length of a measuring rod or the time of a process alters with the motion of the observer. There seems no basis for extending this relativity outside physics to ethics or aesthetics any more than we would let the wave particle duality of quantum theory license a wave particle character for what is morally good. In any case, this sort of relativity is not novel with relativity theory. In classical physics the energy and momentum of an object (and many other quantities) vary with the state of motion of the observer. Relativity theory has just increased the number of quantities with this relative character. Moreover, the emphasis of relativity was an idiosyncrasy nurtured by Einstein. Minkowski (1952, 83) saw the same theory quite differently. He deemed the name "relativity postulate" as a "very feeble" way to label the relevant invariance of the theory and preferred the alternative "postulate of the absolute world" in deference to the entanglement of space and time into a single spacetime. Had the sloganeers attended more closely to Minkowski, might we instead be seeking to deflate the slogan "All is absolute"?

5.2. The Relativity of Motion

Einstein discovered the special theory of relativity when seeking to reconcile the experimentally inviolate relativity of inertial motion with Maxwell's theory of electrodynamics. He then found the general theory as part of his efforts to extend this relativity of inertial motion to accelerated motion. While his motive was clear, it remains unclear whether his general theory does extend the relativity of motion to acceleration. The relativity of inertial motion of special relativity is expressed geometrically in a perfect homogeneity of its spacetime, the Minkowski spacetime. It is exactly analogous to the homogeneity of a Euclidean surface. The equivalence of all inertial states of motion is the analog of the equivalence of all directions in the Euclidean surface. In general relativity the spacetime loses it homogeneity, as does a geometric surface when it adopts varying curvature, such as the variegated surface of a mountain. We can now pick out preferred directions in this surface of varying curvature by adapting our directions to the curvature of the mountainside. Analogously, the varying curvature of the spacetime of general relativity allows us to pick out preferred states of motion; in a standard big bang cosmology, there is a unique rest state associated with the motion of the galaxies. So superficially we cannot draw the moral of the relativity of motion without violating *robustness*.

The considerations rehearsed above are just introductory flourishes in a debate of great complexity with many ingenious proposals and counterproposals. For an extended survey, see Norton 1993, 1995a.

5.3. *Arbitrariness of Coordinate Systems*

One of Einstein's favored expressions of the extended principle of relativity was his principle of general covariance. It asserts our freedom to use any spacetime coordinate system, just as his principle of relativity of motion had allowed us to use any inertial frame of reference in our physics. One might be tempted to claim this as a moral of relativity theory, especially since it came to prominence through Einstein's general theory and so might be expected to respect *robustness*. It may even have an ontological character in so far as it asserts the insubstantiality of spacetime coordinate systems. These expectations fail, however.

Einstein's general theory of relativity was the first prominent spacetime theory to employ arbitrary coordinate systems. There is no simple way of formulating the theory without them. Earlier theories of space and time could also be written in arbitrary coordinate systems, although this possibility was obscured by the fact that the theories could be expressed in especially simple forms in specialized coordinate systems.

Since all spacetime theories admit formulations that use arbitrary coordinate systems, this purported moral violates *novelty*. Since a coordinate system is just a continuous labeling of events with real numbers, we might well wonder how any physical theory could restrict our purely conventional decisions on how we would like to name events. Qualms such as these support the claim, first developed systematically by Erich Kretschmann (1917), that Einstein's principle of general covariance is physically vacuous (see Norton 1993, 1995a).

5.4. Relativity of Geometry

Both Einstein (1954) and Reichenbach, one of his earliest and best-known philosophical interpreters, advocated what we would now call a conventionality of geometry. Calling it the relativity of geometry, Reichenbach (1958, sec. 8) argues that the geometry of a physical space depends on a choice on how lengths are compared in different parts of space. The conventionality of the geometry arises from the convention inherent in this last choice.

We cannot accept this claim as a moral of relativity theory on pain of violation of *novelty*. Nothing in Einstein's or Reichenbach's arguments depends on relativity theory; their arguments can be mounted equally in classical theories. Indeed Henri Poincaré, as both Einstein and Reichenbach acknowledge, had already advocated a version of this conventionality in the form of the claimed conventionality of choice between the geometries of constant curvature (see Friedman 2002). I also remain unconvinced that this conventionality is supportable. If the arguments of Einstein and Reichanbach that support it are acceptable, then it seems to me that we must conclude that anything that is not immediately measurable is also conventional (see Norton 1999b, sec. 5.2).

5.5. Relational View of Space and Time

Einstein presented his theories of relativity as a part of the relational tradition in theories of space and time. That tradition looks upon space and time as some sort of a construct. The real lies in spatial and temporal relations between bodies; space and time are abstractions from those relations. Or the real lies in relations between events; spacetime is an abstraction from them. In the light of the requirement of *realism*, the advent of general relativity would seem not to favor the relationist view. Under a literal reading, general relativity is the theory of a spacetime as a fundamental entity in its own right; it is what endows events with their relational properties, such as the spatial and temporal distances between us. However, too strict a realist reading of general relativity can cause trouble, as we saw in the context of the hole argument above. So we might retreat somewhat from the strongest realist reading. However, that retreat is still far from what a relationist needs. To extract a relationist moral from relativity theory still seems to extract more that can be read uncontroversially in the theory. It seems to violate *modesty* (see Earman 1989).

The most energetically developed relational approach lies in the tradition of Machian theories. Einstein originally saw his general theory of relativity as implementing a demand he saw in the writings of Mach: the inertial properties of a body do not derive from spacetime but from an interaction with all other bodies in the universe. In spite of his early enthusiasm, Einstein came to abandon the demand that his theory of gravity satisfy this requirement. There is a flourishing tradition in Machian theories. Since it generally seeks to augment Einstein's theories in order to realize its brand of relationism, its Machian inspiration cannot be admitted as a moral of relativity theory on pain of violation of *modesty* (see Barbour and Pfister 1995).

6. CONCLUSION

The advent of the relativity theories unsettled and energized philosophy of space and time. In the enthusiasm that followed, it was easy to lose sight of the philosophical morals that were properly to be learned from the relativity theory. They were readily confused with theses that could equally have been advanced and supported prior to Einstein's theories, or those that were appropriate only at an intermediate stage of the development of the theories, or those that Einstein himself found attractive and heuristically useful in his work even though they failed to be implemented in his celebrated discoveries. Yet Einstein's endorsement became as sought after as did Newton's in his time. He lamented, "To punish me for my contempt for authority, Fate made me an authority myself" (quoted in Calaprice 1996, 8).[15]

APPENDIX: A ROBUST VERSION OF THE RELATIVITY OF SIMULTANEITY IN THE MINI-SPACETIMES

The entanglement of measured times and spaces described in section 2 is sufficient to return a version of the relativity of simultaneity that is robust as long as we remain in the mini-spacetimes. This is important, since it shows that the entanglement has captured whatever is essential to the relativity of simultaneity. We can generate this version of the relativity of simultaneity by replicating Einstein's procedure of 1905 in the mini-spacetime. Einstein's procedure was based on a simple definition, illustrated in the spacetime of figure 7.13. We have two positions A and B in space. We send a light signal from A to B, and it is immediately reflected back to A. By Einstein's definition, the event B_1 of the reflection at B is simultaneous with an event A_2 temporally halfway between the emission and the reception of the light signal at A.

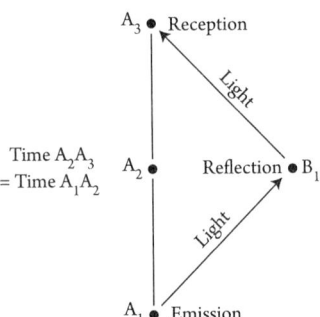

Figure 7.13. Events A_2 and B_1 are simultaneous by Einstein's definition.

15. Attributed to Einstein in Calaprice 1996, 8, where the remark is identified as "Aphorism for a friend, September 18, 1930; Einstein Archive 36–598."

The definition does not depend on light being used for the signal sent from A to B. It gives the same results with any signal as long as we are assured that the speed of the signal in each direction is the same; that is, it takes the same time to go from A to B as from B to A. We will use this relaxed definition below.

...in a Classical Spacetime

If we replicate Einstein's procedure in a classical spacetime, we immediately recover the result that two events, simultaneous for one inertial observer, will be simultaneous for all. Figure 7.14 shows an inertial observer following trajectory $A_1A_2A_3$ with A_2 the event at the temporal midpoint. Let us suppose that we have found an event B_1 that the inertial observer judges as simultaneous with A_2. That means that the two transit times of the signals locating B_1 are equal.

Now consider a second inertial observer who moves in the direction A_2B_1 relative to the first. That observer's trajectory is $A'_1A_2A'_3$, where $A_3A'_3$, A_2B_1 and A'_1A_1 are all parallel. We assume signals A'_1B_1 and $B_1A'_3$ are used to locate B_1. We now repeatedly invoke the lack of entanglement of elapsed times for classical spacetimes illustrated in figure 7.4. That lack of entanglement will assure us that the same time passes for all

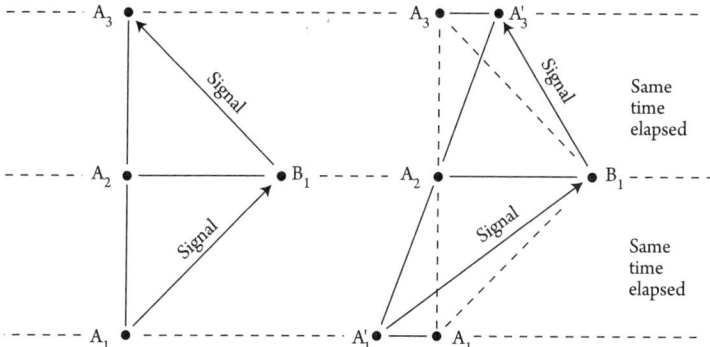

Figure 7.14. In a classical spacetime, if one inertial observer judges events A_2 and B_1 to be simultaneous, then so will all other inertial observers.

the time intervals in the top half of the figure and similarly for the bottom. First we find from it that the time for A'_1A_2 and for A_1A_2 are equal, as are those for A_2A_3 and $A_2A'_3$. Hence all four of these times are equal. By continuing in this way, we quickly find that the transit times for the two new signals A'_1B_1 and $B_1A'_3$ are equal. The conditions of Einstein's revised definition are met, and the new observer judges A_2 and B_1 to be simultaneous.

... in a Relativistic Spacetime

Once we take into account the entanglement of measured spaces and times of relativistic spacetimes, we find that this simple classical result about simultaneity fails. Consider again the two observers $A_1A_2A_3$ and $A'_1A_2A'_3$ as shown in figure 7.14. As before, we locate the event B_1 by requiring that the transit times of the signals A_1B_1 and B_1A_3 are equal. The entanglement of measured times and spaces of figure 7.4 now precludes the same time elapsing along the many intervals shown in figure 7.14, unlike the classical case. In particular it turns out that the transit times for the signals A'_1B_1 and $B_1A'_3$ reflected at event B_1 are unequal even though A_2 is the temporal midpoint of $A'_1A_2A'_3$. The result is that the new observer does not judge events A_2 and B_1 to be simultaneous, unlike the original observer. The new observer must select a new event B'_1 as shown in figure 7.15 to satisfy the requirement that the signal transit times $A'_1B'_1$ and $B'_1A'_3$ are the same.[16]

---

16. If a timelike vector **t** is orthogonal to a spacelike vector **s** so that $g(\mathbf{s},\mathbf{t}) = 0$, then a distinct timelike vector **t'** will not in general also be orthogonal to **s**. This variability is the robust form of the relativity of simultaneity in the tangent space. To see how it arises in the construction of figures 7.14 and 7.15, let the two observer vectors A_1A_2 and A_2A_3 be the same timelike vector **t**. Let the vector indicating simultaneous spacelike separation A_2B_1 be **s**. The two signals A_1B_1 and B_1A_3 are $\mathbf{t} + \mathbf{s}$ and $\mathbf{t} - \mathbf{s}$. The condition that the times elapsed along both signals are the same is

$$g(t+s, t+s) = g(t-s, t-s)$$

Using the linearity of g, this equality becomes

$$g(t,t) + 2g(s,t) + g(s,s) = g(t,t) - 2g(s,t) + g(s,s)$$

ONTOLOGY OF SPACE AND TIME

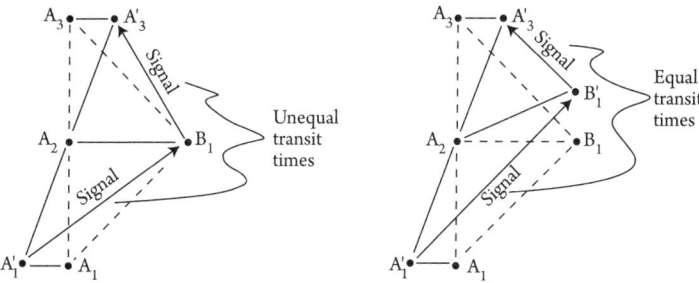

Figure 7.15. In relativistic spacetimes, different inertial observers can disagree on which pairs of event are simultaneous.

This version of the relativity of simultaneity survives only as long as we remain in the mini-spacetimes. Once we relate these mini-spacetimes to the larger spacetime, the richer structure of the larger spacetime may admit a preferred simultaneity relation. To use the earlier example, the preferred simultaneity relation of a Robertson-Walker spacetime can be projected into the mini-spacetime. So this version of the relativity of simultaneity is not admissible as a moral that must respect *robustness*. The entanglement of space and time shown in figure 7.4 does survive when we relate the mini-spacetimes to the larger spacetime. Indeed the entanglement becomes of great importance. Through it we are able to say that free fall trajectories are those along which the maximum time elapses, and this condition can be used as a definition of free fall trajectories. Since classical spacetimes do not have the same entanglement of space and time, no comparable definition is possible in them.

References

Arntzenius, Frank, and Tim Maudlin. 2000. "Time Travel and Modern Physics." In *The Stanford Encyclopedia of Philosophy*, ed. Edward N. Zalta, http://plato.stanford.edu/entries/time-travel-phys/.

---

which is satisfied when $g(\mathbf{s},\mathbf{t}) = 0$; that is, when $\mathbf{s}$ and $\mathbf{t}$ are orthogonal. The above demonstration does not require the signals to be timelike vectors. They could be lightlike, in analogy with Einstein's original derivation, or even spacelike (tachyonic).

Barbour, Julian, and Herbert Pfister, eds. 1995. *Mach's Principle: From Newton's Bucket to Quantum Gravity*. Einstein Studies 6. Boston: Birkhäuser.

Calaprice, Alice, ed. 1996. *The Quotable Einstein*. Princeton, NJ: Princeton University Press.

Čapek, Milič. 1976. "The Inclusion of Becoming in the Physical World." In *The Concepts of Space and Time*, ed. Milič Čapek, 501–524. Boston Studies in the Philosophy of Science 22. Dordrecht, Netherlands: Reidel.

Earman, John. 1989. *World Enough and Spacetime: Absolute versus Relational Theories of Space and Time*. Cambridge, MA: MIT Press.

Earman, John. 1995. *Bangs, Crunches, Whimpers, and Shrieks*. Oxford: Oxford University Press.

Earman, John, and John D. Norton. 1987. "What Price Spacetime Substantivalism? The Hole Story." *British Journal for the Philosophy of Science* 38:515–525.

Earman, John, and John D. Norton. 1993. "Forever Is a Day: Supertasks in Pitowsky and Malament-Hogarth Spacetimes." *Philosophy of Science* 60:22–42.

Einstein, Albert. 1907. "Über das Relativitätsprinzip und die aus demselben gezogenen Folgerungen." *Jahrbuch der Radioaktivität und Elektronik* 4:411–462.

Einstein, Albert. [1905] 1952a. "Zur Elektrodynamik bewegter Körper," *Annalen der Physik*, 17, 891–921; translated as "On the Electrodynamics of Moving Bodies." In *The Principle of Relativity*, by H. A. Lorentz, Albert Einstein, H. Minkowski, and H. Weyl, 37–65. New York: Dover.

Einstein, Albert. [1911] 1952b. "Über den Einfluss der Schwerkraft auf die Ausbreitung des Lichtes," *Annalen der Physik*, 35, 898–908; translated as "On the Influence of Gravitation on the Propagation of Light." In *The Principle of Relativity*, by H. A. Lorentz, Albert Einstein, H. Minkowski, and H. Weyl, 99–108. New York: Dover.

Einstein, Albert. 1954. "Geometry and Experience." In *Ideas and Opinions*, ed. Carl Seelig, 232–246. New York: Crown.

Einstein, Albert. 1956. *The Meaning of Relativity*. Princeton, NJ: Princeton University Press.

Einstein, Albert, and Adriaan D. Fokker. 1914. "Die Nordströmsche Gravitationstheorie vom Standpunkt des absoluten Differentialkalküls." *Annalen der Physik* 44:321–328.

Friedman, Michael. 2002. "Geometry as a Branch of Physics: Background and Context for Einstein's 'Geometry and Experience.'" In *Reading Natural Philosophy: Essays in the History and Philosophy of Science and Mathematics*, ed. David B. Malament, 193–229. Chicago: Open Court.

Grünbaum, Adolf. 1976. "The Exclusion of Becoming from the Physical World." In *The Concepts of Space and Time*, ed. Milič Čapek, 471–500. Boston Studies in the Philosophy of Science 22. Dordrecht, Netherlands: Reidel.

Grünbaum, Adolf. 2001. "David Malament and the Conventionality of Simultaneity: A Reply," http://philsci-archive.pitt.edu/184/.
Hawking, Stephen W., and George F. R. Ellis. 1973. *The Large Scale Structure of Space-Time*. Cambridge: Cambridge University Press.
Hoefer, Carl. 2000. "Energy Conservation in GTR." *Studies in History and Philosophy of Modern Physics* 31:187–199.
Janis, Allen. 2002. "Space and Time, Conventionality of Simultaneity." In *The Stanford Encyclopedia of Philosophy*, ed. Edward N. Zalta, http://plato.stanford.edu/entries/spacetime-convensimul/.
Kretschmann, Erich. 1917. "Über den physikalischen Sinn der Relativitätspostulat, A Einsteins neue und seine ursprünglische Relativitätstheorie." *Annalen der Physik* 53:575–614.
Maxwell, Nicholas. 1993. "Discussion: On Relativity Theory and Openness of the Future." *Philosophy of Science* 60:341–348.
Minkowski, Hermann. 1908. "Raum und Zeit." *Physikalische Zeitschrift* 10:104–111.
Minkowski, Hermann. 1952. "Space and Time." In *The Principle of Relativity*, by H. A. Lorentz, Albert Einstein, Hermann Minkowski, and H. Weyl, 75–91. New York: Dover.
Misner, Charles W., Kip S. Thorne, and John A. Wheeler. 1973. *Gravitation*. San Francisco: Freeman.
Norton, John D. 1985. "What Was Einstein's Principle of Equivalence?" *Studies in History and Philosophy of Science* 16:203–246.
Norton, John D. 1993. "General Covariance and the Foundations of General Relativity: Eight Decades of Dispute." *Reports on Progress in Physics* 56:791–858.
Norton, John D. 1995a. "Did Einstein Stumble: The Debate over General Covariance." *Erkenntnis* 42:223–245.
Norton, John D. 1995b. "Mach's Principle before Einstein." In *Mach's Principle: From Newton's Bucket to Quantum Gravity*, ed. Julian B. Barbour and Herbert Pfister, 9–57. Einstein Studies 6. Boston: Birkhäuser.
Norton, John D. 1999a. "The Hole Argument." In *The Stanford Encyclopedia of Philosophy*, ed. Edward N. Zalta, http://plato.stanford.edu/entries/spacetime-holearg/.
Norton, John D. 1999b. "Philosophy of Space and Time." In *Introduction to the Philosophy of Science*, by Merrilee H. Salmon, John Earman, Clark Glymour, James G. Lennox, J. E. McGuire, Peter Machamer, John D. Norton, Wesley C. Salmon, and Kenneth F. Schaffner, 179–231. Indianapolis: Hackett.
Reichenbach, Hans. 1956. *The Direction of Time*. Berkeley: University of California Press.
Reichenbach, Hans. 1958. *The Philosophy of Space and Time*. Translated by Maria Reichenbach and John Freund. New York: Dover.

Sklar, Lawrence. 1976. *Space, Time, and Spacetime*. Berkeley: University of California Press.

Sklar, Lawrence. 1985. *Philosophy and Spacetime Physics*. Berkeley: University of California Press.

Stachel, John. 1989. "Einstein's Search for General Covariance." In *Einstein and the History of General Relativity*, ed. Don Howard and John Stachel, 63–100. Einstein Studies 1. Boston: Birkhäuser.

Stachel, John. 1993. "The Meaning of General Covariance: The Hole Story." In *Philosophical Problems of the Internal and External World: Essays on the Philosophy of Adolf Grünbaum*, ed. John Earman, Allen I. Janis, G. J. Massey, and N. Rescher, 129–160. Pittsburgh: University of Pittsburgh Press.

Stein, Howard. 1991. "On Relativity Theory and Openness of the Future." *Philosophy of Science* 58:147–167.

Wald, Robert. 1984. *General Relativity*. Chicago: University of Chicago Press

Chapter 8

QM$_\infty$

LAURA RUETSCHE

1. INTRODUCTION

What I will call theories of *ordinary quantum mechanics* (QM) concern systems of finitely many particles. There is a standard notion of what it takes to quantize such a system. One finds a *Hilbert space representation*, that is, a set of operators acting on a Hilbert space and obeying relations characteristic of the system at hand: *canonical commutation relations* (CCRs) for mechanical systems or *canonical anticommutation relations* (CARs) for spin systems. Observables pertaining to the system are obtained by taking linear combinations of and limits of linear combinations of the representation-bearing operators. (This procedure yields what is known as the CCR/CAR *algebra*.) A state of the system is a well-behaved expectation value assignment to these observables. Thus a Hilbert space representation of the CCRs or CARs appropriate to a quantum system supplies a *kinematics*—an account of the possible states and the magnitudes in their scope—for that system.

Most interpreters of ordinary QM take quantum kinematics, in the form determined by a Hilbert space representation, as their point of departure. After that they typically diverge. They disagree about whether to supplement these kinematics' bare quantum states with hidden variables. They disagree about whether quantum *dynamics* (that is, the time evolution of quantum states and observables) is collapse ridden. They disagree about whether a quantum system can exhibit a determinate observable value its state cannot predict with certainty.[1]

Interpreters of ordinary QM have a lot to be anxious about. But they generally need not worry that the structure framing interpretive disputes of the sort just cataloged might admit multiple, *physically inequivalent* instantiations. Suppose there were manifold realizations of the directive "quantize this system" that could not be understood as notational variants on one another. Anxiety, anxiety that there is no univocal quantum theory of the system for interpreters to disagree about to begin with, looms.

But thanks to the Stone–von Neumann uniqueness theorem, not over interpreters of ordinary QM. The theorem states that all Hilbert space representations of the CCRs for a particular classical Hamiltonian theory of finitely many particles stand to one another in a mathematical relation called *unitary equivalence*. Unitary equivalence is widely accepted as a standard of physical equivalence for Hilbert space representations. It follows that all these representations are simply and unalarmingly different ways of expressing the same quantum kinematics. For finitely many spin systems subject to CARs, Wigner's theorem likewise guarantees uniqueness. For systems of finitely many particles, the directive "quantize!" has a unique outcome.

These uniqueness theorems do not extend to quantum field theories (QFTs), which one obtains not by quantizing a system of finitely many particles but by quantizing a field, an entity defined at every point in space (time). Neither do they apply to the thermodynamic limit of quantum statistical mechanics (QSM), where the number of systems one considers and the volume they occupy are taken to be infinite. I will collect such theories under the heading "QM_∞." What is provoking about QM_∞ is this: according to very same criterion of physical equivalence by whose lights Hilbert space representations for ordinary quantum theories are reassuringly unique, a QM_∞ theory can admit infinitely many presumptively physically inequivalent Hilbert space representations. Interpreters of QM_∞ are prone to anxieties unknown to interpreters of ordinary QM.

---

1. Albert 1994 is an accessible overview of some of these disagreements and how they interact.

This chapter chronicles those anxieties. It opens with brief accounts of quantization (section 2) and its uniqueness for ordinary QM (section 3). After illustrating the nonuniqueness of Hilbert space representations in QM_∞ (section 4) and some ensuing interpretive questions (section 5), it sketches the rudiments of the algebraic approach to quantum theories, which discloses a structure common even to unitarily inequivalent Hilbert space representations (section 6). The final section articulates and evaluates different strategies for interpreting QM_∞ in the face of the nonuniqueness of Hilbert space representations. The chapter will be mathematically informal. It will also be limited in scope. Excellent correctives to both these failings are Halvorson and Müger 2007 for QFT[2] and Sewell 2002 for QSM. Ruetsche 2011 elaborates many of the themes introduced here.

2. QUANTIZING

2.1. Representing the CCRs

In classical Hamiltonian mechanics, the state of a simple mechanical system is given by its position and momentum. The components q_i and p_j of the position and momentum variables therefore serve as coordinates for the phase space M of possible states of the system. They are known as *canonical coordinates*. System observables are functions from M to $\mathbb{R}$. The position and momentum *observables*, for example, map points in phase space to their q_i and p_j coordinate values, respectively. All other observables can be expressed as functions of these observables. Of particular significance is the Hamiltonian observable H, which usually coincides with the sum of the system's kinetic and potential energies. The Hamiltonian helps identify dynamically possible trajectories $\mathbf{q}(t)$, $\mathbf{p}(t)$ through phase space as those obedient to William Rowan Hamilton's equations of motion, which are equivalent to Isaac Newton's second law:

$$\frac{dq_i}{dt} = \frac{\partial H}{\partial p_i}, \quad \frac{dp_i}{dt} = -\frac{\partial H}{\partial q_i} \tag{1}$$

---

[2]. See also Huggett 2000; Sklar 2002; and Teller 1995 for philosophical discussions of the important topic, neglected here, of the renormalization of interacting QFTs.

Here and for the sake of mathematical tractability we will confine our attention to linear dynamical systems, that is (more or less) those whose equations of motion are linear in the canonical coordinates.[3]

To prepare the ground for the quantization of a classical Hamiltonian theory, we ought to say a little more. We ought to say something about the *algebraic* structure of classical observables. As smooth functions on phase space, classical observables form a set that is also a vector space over the real numbers. To first approximation, an algebra is a linear vector space V endowed with a (not necessarily associative) multiplicative structure. The vector space of classical observables becomes a *Lie algebra* upon being equipped with a multiplicative structure supplied by the Poisson bracket. The Poisson bracket $\{f, g\}$ of classical observables $f: M \to \mathbb{R}$ and $g: M \to \mathbb{R}$ is

$$\{f,g\} := \sum_i \left(\frac{\partial f}{\partial q_i} \frac{\partial g}{\partial p_i} - \frac{\partial f}{\partial p_i} \frac{\partial g}{\partial q_i} \right) \tag{2}$$

It provides another way of formulating Hamilton's equations (1).

$$\frac{dq_i}{dt} = \{q_i, H\}, \quad \frac{dp_i}{dt} = \{p_i, H\} \tag{3}$$

Notice that for observables p and q

$$\{p_i, p_j\} = \{q_i, q_j\} = 0, \{p_i, q_j\} = -\delta_{ij} \tag{4}$$

Considered both as coordinates and as observables, p_j and q_i are canonical *because* they satisfy (4).

Now to quantize a classical theory cast in Hamiltonian form, one promotes its canonical observables to symmetric operators $\hat{q}_i$, $\hat{p}_i$ acting on a separable[4] Hilbert space $\mathcal{H}$ and obeying commutation relations corresponding to the Poisson brackets of the classical theory. For a mercifully simple classical theory with phase space $\mathbb{R}^{2n}$ and canonical

---

3. For a precise characterization, see Marsden and Raițu 1994, sec. 2.7ff.; Wald 1994, 14ff.
4. Recall that a separable Hilbert space is one with a countable basis.

observables q_i and p_j, these CCRs are (where $[\hat{A}, \hat{B}] := \hat{A}\hat{B} - \hat{B}\hat{A}$ and $\hat{I}$ is the identity operator) (cf. (4))

$$[\hat{p}_i, \hat{p}_j] = [\hat{q}_i, \hat{q}_j] = 0, \quad [\hat{p}_i, \hat{q}_j] = -\frac{\hat{I}\delta_{ij}}{i} \tag{5}$$

2.2. Representing the CARs

To quantize a single spin system, one finds symmetric operators $\{\hat{\sigma}_x, \hat{\sigma}_y, \hat{\sigma}_z\}$ acting on a Hilbert space $\mathcal{H}$ to satisfy the CARs,

$$[\hat{\sigma}_x, \hat{\sigma}_y] = i\hat{\sigma}_z, \quad [\hat{\sigma}_y, \hat{\sigma}_z] = i\hat{\sigma}_x, \quad [\hat{\sigma}_z, \hat{\sigma}_x] = i\hat{\sigma}_y \tag{6}$$

$$\hat{\sigma} \cdot \hat{\sigma} = (\hat{\sigma}_x)^2 + (\hat{\sigma}_y)^2 + (\hat{\sigma}_z)^2 = 3\hat{I} \tag{7}$$

Call these $\{\hat{\sigma}_i\}$ the Pauli spin observables. The generalization to n spin systems is straightforward. To quantize such a system, one finds for each spin system k a Pauli spin $\hat{\sigma}^k = (\hat{\sigma}_x^k, \hat{\sigma}_y^k, \hat{\sigma}_z^k)$ satisfying the CARs, expanded to include the requirement that spin observables for different systems commute:

$$[\hat{\sigma}_x^k, \hat{\sigma}_y^{k'}] = i\delta_{kk'}\hat{\sigma}_z^k, \text{ etc.,} \tag{8}$$

Having crafted a Hilbert space representation of the CARs/CCRs, one prosecutes QM as usual. One enriches one's set of physical magnitudes by taking linear combinations and limits of linear combinations of the canonical magnitudes representing the CARs/CCRs. The result is the self-adjoint part of $\mathfrak{B}(\mathcal{H})$, the bounded operators on $\mathcal{H}$, the Hilbert space carrying the representation.

3. UNIQUENESS IN ORDINARY QM

Werner and Irwin are instructed to quantize a system obeying CCRs and possessing n degrees of freedom. Paul and Wolfgang are asked to do the same for a CAR system of m degrees of freedom. No matter how idiosyncratically Werner and Irwin proceed, they produce the same quantization—or so the Stone–von Neumann theorem is generally understood to imply. No matter how perversely Paul and Wolfgang proceed, their quantizations coincide as well—or so Wigner's theorem is generally understood to imply. Getting a handle on how these implications work is a useful a prolegomenon to interpreting QM_∞.

Think of a physical theory as running a sort of modal toggle. The theory sorts logically possible worlds into two sorts: those that are (by its lights) physically possible and those that are not. The content of a theory so construed consists of the set of worlds possible according to it. And two theories are physically equivalent—they have the same content—just in case they put the same worlds into the "physically possible" pile. This content coincidence criterion of physical equivalence, which descends from the modal toggle picture of a physical theory, is one part of the background against which the Stone–von Neumann and Wigner theorems emerge as demonstrations of the physical equivalence of quantizations in their scope.

Another part of the background is a set of substantive assumptions about the particular shape of those quantizations and in particular about the structures with respect to which they describe and individuate their possible worlds. A quantization is modeled as a pair $(\mathcal{M}, \mathcal{S})$ whose first entry denotes the set of physical magnitudes it recognizes and whose second entry denotes its set of states. Given that a state is a map from magnitudes to their expectation values, an ideology about how such maps should behave could be sufficient to determine $\mathcal{S}$ given $\mathcal{M}$. For instance, a theory of ordinary QM equates $\mathcal{M}$ with $\mathfrak{B}(\mathcal{H})$ for some separable $\mathcal{H}$ and adopts the ideology that states should be normalized, linear, positive, and countably additive. It follows from Gleason's theorem that ordinary QM must equate $\mathcal{S}$ with $\mathfrak{T}^+(\mathcal{H})$, the set of density operators on $\mathcal{H}$. Each $\widehat{W} \in \mathfrak{T}^+(\mathcal{H})$ determines a state via the trace prescription, which assigns each self-adjoint $\hat{A} \in \mathfrak{B}(\mathcal{H})$ the expectation value $Tr(\widehat{W}\hat{A})$. I persist in modeling a quantization as an ordered *pair* $(\mathcal{M}, \mathcal{S})$, because ideologies of state are discretionary, even optional. Imposing a *universal* rule for generating $\mathcal{S}$ given $\mathcal{M}$ would obliterate a certain sort of interpretational freedom.

The pair $(\mathfrak{B}(\mathcal{H}), \mathfrak{T}^+(\mathcal{H}))$ encapsulates what worlds are possible according to a quantization of the sort under discussion. The sets of worlds possible according to quantizations $(\mathfrak{B}(\mathcal{H}), \mathfrak{T}^+(\mathcal{H}))$ and $(\mathfrak{B}(\mathcal{H}'), \mathfrak{T}^+(\mathcal{H}'))$ coincide exactly when there are representation-preserving isomorphisms $i_{obs} : \mathfrak{B}(\mathcal{H}) \to \mathfrak{B}(\mathcal{H}')$ and $i_{state} : \mathfrak{T}^+(\mathcal{H}) \to \mathfrak{T}^+(\mathcal{H}')$ such that

$$Tr\left(i_{state}(\widehat{W})i_{obs}(\widehat{A})\right) = Tr(\widehat{W}\widehat{A}) \qquad (9)$$

for all $\widehat{W} \in \mathfrak{T}^+(\mathcal{H})$ and all $\widehat{A} \in \mathfrak{B}(\mathcal{H})$.[5] (9) ensures that to every state $\widehat{W}$ in one quantization there stands a state $i_{state}(\widehat{W})$ in the other that is the counterpart of the first in the sense that $i_{state}(\widehat{W})$'s assignment of expectation values to observables $i_{obs}(\widehat{A})$ perfectly mimics $\widehat{W}$ assignment of expectation values to observables $\widehat{A}$. We are a definition away from being able to announce a necessary and sufficient condition for the physical equivalence of quantizations.

A Hilbert space $\mathcal{H}$ and a collection of operators $\{\widehat{O}'_i\}$ is *unitarily equivalent* to $(\mathcal{H}', \{\widehat{O}'_i\})$ if and only if there exists a one-to-one, linear, norm-preserving transformation ("unitary map") $U : \mathcal{H} \to \mathcal{H}'$ such that $U^{-1}\widehat{O}'_i U = \widehat{O}_i$ for all i.

Now for the announcement: quantizations $(\mathfrak{B}(\mathcal{H}), \mathfrak{T}^+(\mathcal{H}))$ and $(\mathfrak{B}(\mathcal{H}'), \mathfrak{T}^+(\mathcal{H}'))$ are physically equivalent if and only if $(\mathcal{H}, \mathfrak{B}(\mathcal{H}))$ and $(\mathcal{H}', \mathfrak{B}(\mathcal{H}'))$ are unitarily equivalent. The U effecting their unitary equivalence furnishes both the bijection i_{state} from the first theory's state space to the second's and the bijection i_{obs} from the first theory's observable set to the second's. Unitary equivalence is the relation, the Stone–von Neumann theorem assures us, all representations of the CCRs for n degrees of freedom (n finite) enjoy with one another.[6] It is the relation, Wigner's theorem assures us, all representations of the CARs for n degrees of freedom (n finite) enjoy with one another. According to the content coincidence criterion of physical equivalence, unitarily equivalent quantizations are physically equivalent.

---

5. Clifton and Halvorson 2001 develops and deploys this criterion of physical equivalence.
6. This needs to be qualified slightly. See Summers 2001 for details.

4. NONUNIQUENESS IN QM_∞

QM_∞ abounds with what the Stone–von Neumann and Wigner theorems declared unimaginable for ordinary QM: unitarily inequivalent representations of the CCR/CAR algebras. Anyone even minimally acquainted with philosophical treatments of QM has on hand the resources to describe a quantum system admitting unitarily inequivalent representations: a chain of infinitely many spin $\frac{1}{2}$ systems.[7]

As a warm-up, consider a finite number of spin $\frac{1}{2}$ systems arranged in a one-dimensional lattice. The CARs ([6])–[8]) specify what it takes to quantize this system. One way to build a Hilbert space representation fitting these specifications is to use a vector space $\mathcal{H}$ spanned by a basis consisting of sequences s_k, where k ranges from $-n$ to n, and each entry takes one of the values ± 1. (NB there are finitely many distinct such sequences—finitely many ways to map a set of finite cardinality into the set $\{+1, -1\}$.) Operators $\hat{\sigma}_z^m$ are introduced in such a way that sequences s_k whose m^{th} entry is ± 1 correspond to eigenvectors associated with the eigenvalue ± 1. Operators $\hat{\sigma}_x^k, \hat{\sigma}_y^k$ conspiring with these to satisfy the CARs are introduced by analogy to their single electron counterparts.

The *polarization* of a system is described by a vector whose magnitude ($\in [0, 1]$) gives the strength and whose orientation gives the direction of the system's magnetization. On a single electron it is represented by an observable $\hat{m}$ whose three components correspond to three orthogonal components of quantum spin. For example, in the +1 eigenstate $|+\rangle$ of $\hat{\sigma}_z$ (understood as the z-component of spin), $\langle \hat{m} \rangle = +1$ along the z axis. For the finite spin chain, the polarization observable has components $\hat{m}_i := 1/(2n+1)\sum_{k=-n}^{n} \hat{\sigma}_i^k$. $\hat{m}$ belongs to $\mathfrak{B}(\mathcal{H})$, because its components are polynomials of Pauli spins.

In the basis sequence s_k, the polarization observable $\hat{m}$ takes an expectation value of magnitude $1/(2n+1)\sum_{j=-n}^{n}[s_k]_j$ (where $[s_k]_j$ is the jth entry of the sequence s_k) directed along the z axis. The polarization attains extreme values (of ± 1) for those sequences every term of which is the same. Let $\widehat{W}$ be a state in this quantization assigning $\hat{m}_z$ the

---

7. Here I follow Sewell 2002, sec. 2.3, to which I refer the reader for details.

expectation value +1. Wigner's theorem ensures that any other representation of the CARs will be unitarily equivalent to this one. Any other representation of the CARs is thus guaranteed to contain a state $\widehat{W}'$ (the image of $\widehat{W}$ under the unitary map implementing the equivalence of the representations) and an observable $\hat{m}'_z$ (the image of $\hat{m}_z$ under that map) such that the expectation value of $\hat{m}'_z$ in the state $\widehat{W}'$ is +1.

Now consider a doubly infinite chain, labeled by the positive and negative integers $\mathbb{Z} = \{\ldots, -2, -1, 0, 1, 2,\ldots\}$, of spin $\frac{1}{2}$ systems. As before, to quantize this system we must associate with each site k a trio $\hat{\sigma}_k = (\hat{\sigma}^k_x, \hat{\sigma}^k_y, \hat{\sigma}^k_z)$ satisfying the CARs. But if we follow the strategy adopted for the finite spin chain and attempt to construct our Hilbert space from a basis consisting all possible maps from $\mathbb{Z}$ to $\{\pm 1\}$, we are foiled. Because the set of such maps is nondenumerable, the Hilbert space we would construct would be nonseparable, breaking the tradition of using separable Hilbert spaces for physics.[8]

Here is one way to build a separable Hilbert space representation of the CARs for an infinite chain of spins. Start with the sequence $[s_k]_j = +1$ for $j \in \mathbb{Z}$ and add all sequences obtainable therefrom by finitely many local modifications. The resulting basis consists of all sequences for which all but a finite number of sites take the value +1. Continue to follow the model of the finite spin chain to introduce operators $\hat{\sigma}^+_k$ satisfying the CARs. I will call this representation $(\mathcal{H}^+, S^+)$.

Before considering how the polarization observable $\hat{m}^+$ behaves on $(\mathcal{H}^+, S^+)$, let us assure ourselves that there is such an observable. To be an element of $\mathfrak{B}(\mathcal{H}^+)$, the polarization observable must be a polynomial of Pauli spins or the limit (in the appropriate sense) of a sequence of such polynomials. The sequence $\hat{m}^N_i := 1/(2N+1)\Sigma^N_{k=-N} \hat{\sigma}^k_i$ has a limit as $N \to \infty$ in $\mathcal{H}^+$'s weak topology, which provides the appropriate sense of convergence. A sequence $\hat{A}_i$ of operators on $\mathcal{H}$ converges to an operator $\hat{A}$ in $\mathcal{H}$'s weak operator topology if for all $|\psi\rangle, |\phi\rangle \in H, |\langle \psi|(\hat{A} - \hat{A}_i)|\phi\rangle|$ goes to 0 as i goes to ∞. Confining attention to the sequence $\hat{m}^N_z$, observe that $\langle s_k | \hat{m}^N_z | s'_k \rangle = 1/(2N+1)\Sigma^N_{j=-N} [s_k]_j + [s'_k]_j /2$ for basis sequences s_k and s'_k. Because −1 occurs only finitely many times in each basis sequence,

---

8. Halvorson 2004 defends just such iconoclasty.

$\langle s_k | \hat{m}_z^N | s_k' \rangle / (2N+1)$ will converge to 1 as $N \to \infty$, no matter what s_k and s_k' are. This shows that the sequence $\hat{m}_z^N$ converges weakly to the identity operator. We can imagine realizations of the CARs, modeled on $(\mathcal{H}^+, S^+)$, for which this is not so: for example, a representation whose "ground state" s_k is a sequence for which $\lim_{N\to\infty} \langle s_k | \hat{m}_i^N | s_k \rangle$ does not converge (e.g., the sequence $[s_k]_j = +1$ for $2^n < |j| \leq 2^{n+1}$ and n odd; $[s_k]_j = -1$ otherwise).

Other components of global polarization are also captured as weak limits of polynomials of Pauli spins. Each component of polarization is thus an element of $\mathfrak{B}(\mathcal{H}^+)$, and so an observable in ordinary QM's sense. For every state s_k in the basis, $\langle \hat{m}^+ \rangle$ will be oriented along the z axis and take the value $\lim_{n\to\infty} 1/(2n+1) \Sigma_{j=-n}^{n} [s_k]_j$. Because for each basis element, all but a finite number of its entries take the value +1, this limit will be 1. Every state representable on S^+ will inherit this feature from the basis vectors in terms of which it is expressed: every state in the representation will have unit polarization in the positive z direction.

Because the chain is infinite, Wigner's theorem does not imply that the quantization just constructed is unique up to unitary equivalence. And it is not. Consider, for instance, a representation set in a Hilbert space whose basis elements correspond to the sequence $[s_k]_j = -1$ for $j \in \mathbb{Z}$ along with all sequences obtainable from this one by finitely many local modifications. The basis consists, then, of sequences for which all but a finite number of sites take the value -1. Operators $\hat{\sigma}_k^-$ satisfying the CARs are introduced in such a way that the nth entry in the basis sequence s_k gives the expectation value of $\hat{\sigma}_z^n$. Call this representation $(\mathcal{H}^-, S^-)$. By parity of reasoning, $\hat{m}^-$ is an observable in this quantization, one assigned the expectation value -1 by every state in S^-.

We can now see that $(\mathcal{H}^-, S^-)$ and $(\mathcal{H}^+, S^+)$ are not unitarily equivalent. Suppose, for contradiction, that they were. Then there would be a unitary $U : \mathcal{H}^+ \to \mathcal{H}^-$ such that $U\hat{\sigma}_n^+ U^{-1} = \hat{\sigma}_n^-$ for all n. Where $\hat{m}_N^\pm := 1/(2N+1) \Sigma_{n=-N}^{N} \hat{\sigma}_n^\pm$, this implies that $\hat{m}_N^- = U\hat{m}_N^+ U^{-1}$. For $|\psi^+\rangle$ and $|\psi^-\rangle$, unit vectors in $\mathcal{H}^+$ and $\mathcal{H}^-$ related by $|\psi^-\rangle = U|\psi^+\rangle$, it follows that

$$\langle \psi^+ | \hat{m}_N^+ | \psi^+ \rangle = \langle \psi^- | \hat{m}_N^- | \psi^- \rangle \tag{10}$$

But in the limit $N \to \infty$, (10) breaks down: the right-hand side (r.h.s.) (which gives the expectation value the state $|\psi^-\rangle$ assigns the polarization observable $\hat{m}^-$) and the left-hand side (l.h.s.) (which gives the expectation value the state $|\psi^+\rangle$ assigns the polarization observable $\hat{m}^+$) go to +1 and −1, respectively. The expectation values $|\psi^+\rangle$ assigns observables in $\mathfrak{B}(\mathcal{H}^+)$ do not replicate the expectation values $|\psi^-\rangle$ assigns the images of those observables (in $\mathfrak{B}(\mathcal{H}^-)$) under the map U. The quantizations $(\mathcal{H}^-, S^-)$ and $(\mathcal{H}^+, S^+)$ are not unitarily equivalent. We can generate continuously many unitarily inequivalent quantizations of the infinite spin chain by suitable modifications of the strategy used to obtain $(\mathcal{H}^-, S^-)$ and $(\mathcal{H}^+, S^+)$.[9] These quantizations are presumptively physically inequivalent as well.

The surfeit is not peculiar to representations of the CARs. The CCRs for infinitely many degrees of freedom also admit continuously many unitarily inequivalent representations. Indeed the representation (q'_i, p'_i) obtained from a standard representation (q_i, p_i) by the apparently innocuous rescaling

$$p'_i = \frac{p_i}{a}, \; q'_i = aq_i \tag{11}$$

is unitarily inequivalent to its progenitor (Segal 1967).

Unitarily inequivalent representations of the CARs contradict one another concerning the apparently physical matter of the magnitude and direction of the global polarization. There is also something evidently physical at stake between unitarily inequivalent representations of the CCRs. One way to see this is to consider the conventional representation for the CCRs arising from a classical field theory, such as Klein-Gordon theory for a fixed mass m (see Wald 1994, chap. 3). The ensuing quantization tempts a particle interpretation. It incorporates a state $(|0\rangle)$ whose energy and momentum suggest the absence of

---

9. Here, roughly, is why. The strategy was, starting from a particular map s_0 from $\mathbb{Z}$ to $\{\pm 1\}$, to build a vector space from the collection of such maps differing from s_0 in only finitely many places. "Differing from one another in only finitely many places" is an equivalence relation on the set of maps from $\mathbb{Z}$ to $\{\pm 1\}$, which set has the cardinality of the continuum. Each equivalence class has the cardinality of the natural numbers and corresponds to a quantization of the infinite spin chain–different equivalence classes to unitarily inequivalent quantizations. So there must be continuously many inequivalent quantizations.

particles as well as states ("single particle states") whose energies and momenta suggest the presence of one particle and so on. Every state in the representation's Hilbert space can be obtained as a linear combination of states of this sort or as a limit of a sequence of such linear combinations. The quantization moreover hosts an observable $\widehat{N}$ that appears to count particles. $\widehat{N}$'s expectation value in $|0\rangle$ is 0; $\widehat{N}$'s expectation value in an n-particle state is n. Unconventional representations of the CCRs can also have structures inviting a particle interpretation: a vacuum state $|0'\rangle$, a total number operation $\widehat{N}'$, and so on.

Let $(|0\rangle, \widehat{N})$ and $(|0'\rangle, \widehat{N}')$ stand for unitarily inequivalent, particle-interpretation-tempting quantizations. The vacuum state of one is not even a possible state, according to the other; the total number operator of one just is not a physical magnitude, according to the other. Applying to disjoint domains, the particle concepts circumscribed by inequivalent quantizations of the Klein-Gordon field are something like incommensurable. But the quantized Klein-Gordon field was supposed to describe just one uncomplicated kind of particle, the free boson of mass m. Continuously many unitarily—and so, presumptively, physically—inequivalent quantizations of the Klein-Gordon field are an embarrassment of riches that demands further investigation.[10]

Investigations of the formal structure of standard realizations of quantum field theoretic CCRs conducted in the 1950s and 1960s made possible an intriguing commentary on this embarrassment. Each concrete Hilbert space representation of the CCRs gives rise to an algebra that is representation independent. The *Weyl algebra* for a system is constructed by taking linear combinations of operators satisfying the CCRs for that system, then closing in the *norm* topology, whose criterion of convergence is stricter than that of the weak topology we encountered earlier. This Weyl algebra is unique in the sense that no matter what Hilbert space realization of the CCRs for the system one starts with, one obtains the same Weyl algebra. Thus independent of its Hilbert space antecedents, the Weyl algebra for a system can be regarded as an *abstract* algebra, the algebraic structure shared by all its Hilbert space

---

10. For further discussion of inequivalent particle concepts in QFT, see Arageorgis et al. 2003 and Clifton and Halvorson 2001.

representations. For a system whose canonical observables obey CARs, one can likewise abstract from those CARs a single algebraic structure common to every Hilbert space representation of those CARs.

Rudolf Haag and Daniel Kastler (1964, 852) give voice to a suspicion prompted by this circumstance: "The relevant object is the abstract algebra and not the representation. The selection of a particular... representation is a matter of convenience without physical implications." Ordinary QM completes the kinematic template $(\mathcal{M}, \mathcal{S})$ in terms set by a particular concrete Hilbert space $\mathcal{H}$. The suspicion is that *these are the wrong terms* for QM_∞, that a theory of QM_∞ deploys states and observables that *are not* states and observables in ordinary QM's sense. The balance of this chapter articulates and evaluates different strategies for making this suspicion precise.

5. INTERPRETING QM_∞: SOME QUESTIONS

A question looms from the wreckage of the uniqueness theorems: how do theories of QM_∞ characterize and individuate physical possibilities? Understand this as follows: how on behalf of such theories should we complete the template $(\mathcal{M}, \mathcal{S})$? On the answer to this question hinges the answer to another: under what circumstances are theories of QM_∞ physically equivalent? To answer these questions is to begin to interpret QM_∞.

Before canvassing answers, we should consider criteria of adequacy to which we might hold them. These criteria are provisional; they are meant to help initiate, not terminate, the interrogation of interpretive options. The exigencies of interpreting QM_∞ may induce us to revise our sense of what it takes to comprehend the physical world by means of an empirical theory. To my mind, the capacity of interpretive projects to inspire such revision is what makes them philosophical.

In "The Theoretician's Dilemma" Carl Hempel (1965) considers the case for taking the content of a scientific theory to be its "Craigified" part—that is, the connections it draws between observable phenomena, connections presented without the intermediary of (nonlogical) theoretical apparatus. Hempel resists the Craigification of theoretical content, because he attributes theories a function that he considers their

Craigified parts ill-suited to fulfill. That function is explanation. What promotes it, Hempel suggests, is the systematic unification between observable phenomena effected by the theoretical apparatus.

Without going deeper into the details of Hempel's position, we can extract from it one criterion of interpretive adequacy. An interpretation should recognize enough states and enough observables to enable the theory it interprets to discharge its explanatory duties. I do not need—I am not sure there exists—a categorical imperative for identifying these duties. In applying the criterion, we can start from the explanatory aspirations of physicists using a theory and aim for a reflective equilibrium between the assessment of interpretations for sustaining those aspirations and the assessment of those aspirations as appropriate.

"Explanatory duties" is not only vague but also synechdochical. There are a host of things a theory ought to do, and interpretations of a theory are adequate only insofar as they enable the theory to do these things. Because we are approaching the matter of interpretation via the kinematic question of how to fill in the $(\mathcal{M}, \mathcal{S})$ template, it is worth emphasizing that one thing a physical theory ought to do is furnish *dynamics* for the systems it describes. An interpretation that completes the template in a way that hamstrings this dynamical project is on that count suspect.

I have been proceeding as though each theory were an island, complete unto itself. In this mode of address adequacy of the sort under discussion is a purely internal affair: an interpretation ought to attribute a theory sets of observables and magnitudes rich enough to meet that theory's own aims. But externally oriented questions of sufficiency arise as well. We may, for instance, want to interpret a theory in a way that makes sense of its fit with environing theories and their interpretations. Thus in the case of the thermodynamic limit of QSM, we might hope for an interpretation that attributes the theory enough observables/magnitudes to bring it into a substantial explanatory relationship with thermodynamics. Again, existing theories are conceptual resources from which new theories are forged. Thus we might want an interpretation to recognize enough observables/magnitudes to enable an interpreted theory to serve as such a resource. Semiclassical quantum gravity gives a good example of how this desire plays out in the interpretation of QFT.

Semiclassical quantum gravity considers a quantum field on spacetime manifold subject to Albert Einstein's field equations, with a quantum commodity (the expectation value $< T_{ab} >$ of the stress energy tensor T_{ab}) substituted for a classical one (plain old T_{ab}):

$$G_{ab} = 8\pi < T_{ab} > \qquad (12)$$

One of the points of this hybridization is the clues it might hold for what a purebred quantum theory of gravity would look like. An interpretation of QFT that withheld physical significance from a stress energy observable (or something like it) would blunt that point. It would fail to recognize enough observables to sustain QFT in this developmental task.

One vague and provisional criterion of adequacy for an interpretation of QM_∞ is that it recognize enough states and observables. Another is that it not recognize too many. Although this demand has undeniable intuitive pull, I find it difficult to motivate ecumenically. It is preaching to converted Ockhamists to declare parsimony desirable in itself. Perhaps it helps to observe that parsimony can serve recognized epistemological and metaphysical goals. For instance, there is a failure of parsimony that frustrates the venerable goal of *determinism*. For a theory to have any shot at determinism, its dynamics must specify how the values of some set of magnitudes at one time depend on the values of some set of magnitudes at another. Unparsimoniously recognizing more magnitudes than are dynamically tractable, or more states than can be put into one-to-one correspondence with valuations of tractable dynamical magnitudes, a completion of the kinematic template $(\mathcal{M}, \mathcal{S})$ threatens determinism. Of course other components of an interpretation incorporating such a template could work to defuse this threat; see, for instance, defenses of substantivalism against the hole argument (Brighouse 1994).

6. INTERMISSION: ALGEBRAICA

To better appreciate the suspicion raised at the close of section 4, the algebraic framework that grounds it, and the interpretive options that framework makes available, let us undertake a brief review of the rudiments of algebraic approaches to quantum theories. Intuitively, an *algebra* is just a collection of elements along with a way of taking their products and linear combinations. Insofar as it is the business of natural science to weave physical magnitudes into functional relationships, organizing physical magnitudes into an algebra underwrites that business. Officially, an algebra $\mathfrak{A}$ over the field $\mathbb{C}$ of complex numbers is a set of elements $(A, B, \ldots)$ with the following features:

(i) First, $\mathfrak{A}$ is closed under a commutative, associative operation $+$ of binary addition, with respect to which A forms a group.
(ii) Second, $\mathfrak{A}$ is closed with respect to a binary multiplication operation $\cdot$, which is associative and distributive with respect to addition but not necessarily commutative.
(iii) Finally, $\mathfrak{A}$ is closed with respect to multiplication by complex numbers.

Features (i) and (iii) reveal an algebra to be something well known to those acquainted with ordinary QM—a linear vector spare—equipped by feature (ii) with a binary multiplication operation.

More specialized algebras augment this basic structure. A * algebra, is an algebra closed under an *involution* $^* : \mathfrak{A} \to \mathfrak{A}$.[11] $\mathfrak{B}(\mathcal{H})$ is a * algebra, with the adjoint operation † supplying the involution. In ordinary QM an $\hat{A} \in \mathfrak{B}(\mathcal{H})$ such that $\hat{A}^\dagger = \hat{A}$ is *self-adjoint* and represents an observable magnitude. This is generalized by algebraic QM, which deploys * algebras (without prejudice to whether they coincide with some $\mathfrak{B}(\mathcal{H})$ or not) and identifies their self-adjoint elements with observables.

6.1. C^*-algebras

One sort of interweaving of physical magnitudes constitutes one magnitude—the global polarization of an infinite spin chain, say—as the limit of a

---

11. Satisfying: $(A^*)^* = A$, $(A + B)^* = A^* + B^*$, $(cA)^* = \bar{c}A^*$, $(AB)^* = B^*A^*$ for all $A, B \in \mathfrak{A}$ and all complex c.

sequence of other magnitudes. To determine which sequences of elements of an algebra have limits, we need a criterion of convergence for such sequences. A *norm* underwrites such a criterion. A norm on a space is a function that assigns a nonnegative real number to each element of that space.[12] The *Hilbert space vector norm*, which maps each vector $|\psi\rangle$ in a Hilbert space to $\sqrt{\langle\psi|\psi\rangle}$, is an example. A space equipped with a norm is thereby equipped with the criterion of convergence (or, speaking somewhat loosely, a *topology*) induced by that norm: A sequence of v_i of elements of the space V converges to the element v in the norm $\|\ \|: V \to \mathbb{R}$ if and only if the sequence of real numbers $\|v_n - v\|$ converges to 0 in the usual sense.

We have said enough to abstractly characterize an object of extraordinarily wide application. A C^*-algebra $\mathfrak{A}$ is a * algebra over $\mathbb{C}$, equipped with a norm, satisfying $\|A^*A\| = \|A\|^2$ and $\|AB\| \leq \|A\|\ \|B\|$ for all $A, B \in \mathfrak{A}$, with respect to which norm it is complete.[13] A familiar example of a C^*-algebra is the set $\mathfrak{B}(\mathcal{H})$ of bounded operators on a separable Hilbert space $\mathcal{H}$. The Hilbert space operator norm

$$||A|| := \sup_{\sqrt{\langle\psi|\psi\rangle}=1} |A|\psi\rangle| \tag{13}$$

(where $|\psi\rangle$ ranges over $\mathcal{H}$, $|\ |$ is the vector norm and $\sup_{x \in X} f(x)$ is the least upper bound of the values the function $f(x)$ assumes in the domain X) provides the C^*-algebraic norm. It is no accident that $\mathfrak{B}(\mathcal{H})$ is an example of a C^*-algebra. Concretely characterized, *a C^*-algebra is a norm-closed self-adjoint subalgebra of* $\mathfrak{B}(\mathcal{H})$ *for some* $\mathcal{H}$.

While every $\mathfrak{B}(\mathcal{H})$ is a C^*-algebra, the converse is not the case: a C^*-algebra need only be (up to isomorphism) a *sub*algebra of $\mathfrak{B}(\mathcal{H})$. To see how some elements of $\mathfrak{B}(\mathcal{H})$ could be left out of the norm closure of others, reflect that the norm topology is relatively strict. A Hilbert space sustains other operator topologies with looser admissions criteria. The weak operator topology, encountered in section 4's discussion

---

12. In more detail, a norm on a vector space V is map $\|\ \|$ from V into the nonnegative reals satisfying $\|cv\| = |c|\|v\|$, $\|v\|\ \|w\| = \|vw\|$, and $\|v\| + \|w\| > \|v + w\|$ for all $v, w \in V$ and all $c \in \mathbb{C}$.

13. That is, the limit of every norm-convergent sequence of elements of $\mathfrak{A}$ is itself an element of $\mathfrak{A}$.

of whether there really was a polarization observable for the infinite spin chain, is one example. There are others: the strong topology, the σ-strong topology, and so on (see Bratteli and Robinson 1987, sec. 2.4.1, for a disentanglement). As the names imply, strong convergence implies weak convergence but not vice versa; norm convergence implies strong convergence but not vice versa.

The infinite spin illustrates the asymmetry of these implications. As section 4 indicates, the sequence $\hat{m}_z^N$ has a limit as $N \to \infty$ in $\mathcal{H}^+$'s weak topology. That limit is the identity operator. But the same sequence lacks a norm limit. In order for $\hat{m}_z^N$ to converge to $\hat{I}$ in $\mathcal{H}^+$'s norm topology, it must be the case that

$$\left\| \hat{m}_z^N - \hat{I} \right\| := \sup_{\sqrt{\langle \psi | \psi \rangle} = 1} |(\hat{m}_z^N - \hat{I})|\psi\rangle| \qquad (14)$$

approaches 0 as $N \to \infty$. But this will not happen. For each N one can find a basis sequence s_k—one starting with N 0's—s.t. $1/(2n+1)\Sigma_{k=-N}^{N} \hat{m}_z^k |s_k\rangle = 0$. This implies that $\hat{m}_z^N |s_k\rangle = 0$, and so that $|(\hat{m}_z^N - \hat{I})|s_k\rangle| = 1$, and so that $\sup_{\sqrt{\langle \psi | \psi \rangle} = 1} |(\hat{m}_z^N - \hat{I})|\psi\rangle| \geq 1$ —which spoils norm convergence.

A C^*-algebra that is conceived in abstraction admits a concrete Hilbert space representation. A *representation* of a C^*-algebra $\mathfrak{A}$ is an algebraic-structure-preserving map $\pi : \mathfrak{A} \to \mathfrak{B}(\mathcal{H})$ from the C^*-algebra into (although not necessarily onto) the algebra $\mathfrak{B}(\mathcal{H})$ of bounded linear operators on a Hilbert space $\mathcal{H}$. The notion of unitary equivalence extends straightforwardly to representations of C^*-algebras.

6.2. Von Neumann Algebras

The representation $\pi(\mathfrak{A})$ of a C^*-algebra $\mathfrak{A}$ is closed in the norm topology (Bratteli and Robinson 1987, proposition [prop.] 2.3.1, 42). Thus starting with $\pi(\mathfrak{A})$ and closing in the weaker topologies $\mathcal{H}$ makes available could eventuate in sets of observables richer than those contained in $\pi(\mathfrak{A})$ proper. Next we consider algebras obtained by following a recipe of this sort.

Concretely characterized, a *von Neumann algebra* $\mathfrak{R}$ is a *-algebra of bounded operators that is strong operator closed in its action on some

Hilbert space. If you want a von Neumann algebra, start with some self-adjoint algebra of bounded Hilbert space operators—$\pi(\mathfrak{A})$, say—and close in the strong operator topology.

Given an algebra $\mathfrak{D}$ of bounded operators on a Hilbert space $\mathcal{H}$, its *commutant* $\mathfrak{D}'$ is the set of all bounded operators on $\mathcal{H}$ that commute with every element of $\mathfrak{D}$. The commutant of $\mathfrak{B}(\mathcal{H})$, for example, consists of scalar multiples of the identity operator. Naturally enough, $\mathfrak{R}$'s *double commutant* $\mathfrak{D}''$ is $\mathfrak{D}'$'s commutant. Every element of $\mathfrak{B}(\mathcal{H})$ commutes with the identity operator and so with every element of $\mathfrak{B}(\mathcal{H})'$. This makes $\mathfrak{B}(\mathcal{H})$ its own double commutant.

Here is why that is interesting: von Neumann's celebrated double commutant theorem links the "topological" characterization of a von Neumann algebra in terms of closure with an "algebraic" characterization in terms of commutants. Von Neumann showed that the strong and weak closures of a self-adjoint algebra $\mathfrak{D}$ of bounded Hilbert space operators coincide—and coincide as well with $\mathfrak{D}$'s double commutant. Thus a *von Neumann algebra* $\mathfrak{R}$ is a *-algebra of bounded operators such that $\mathfrak{R} = \mathfrak{R}''$. We have just seen a splendid example: $\mathfrak{B}(\mathcal{H})$. Where π is a representation of a C^*-algebra $\mathfrak{A}$, $\pi(\mathfrak{A})''$—which according to the double commutant theorem is equivalent to $\pi(\mathfrak{A})$'s weak closure—is a von Neumann algebra known as *the von Neumann algebra affiliated with the representation π*.

Representations of $\mathfrak{A}$ are *quasi-equivalent* when their affiliated von Neumann algebras are isomorphic. That is, the representations π and φ are quasi-equivalent if and only if there is a *isomorphism a from $\varphi(\mathfrak{A})''$ to $\pi(\mathfrak{A})''$ such that $a[\varphi(A)] = \pi(A)$ for all $A \in \mathfrak{A}$. Unitarily equivalent representations are quasi-equivalent: the *isomorphism is implemented unitarily.

6.3. Algebraic States and GNS Representations

The algebraic approach to quantum theories associates observables pertaining to a system with the self-adjoint elements of a C^*-algebra $\mathfrak{A}$ appropriate to the system. For instance, $\mathfrak{A}$ could be the algebra that results from the norm closure of a set of operators satisfying the CCRs/CARs for the system. The algebraic approach accommodates ordinary

QM's picture of observables as a special case: $\mathfrak{A}$ is set equal to $\mathfrak{B}(\mathcal{H})$ for some Hilbert space $\mathcal{H}$.

Ordinary QM conceives states on $\mathfrak{B}(\mathcal{H})$ as normed, positive, countably additive linear functionals over that algebra. Provided $dim(\mathcal{H}) > 2$, Gleason's theorem places states, so conceived, in one-to-one correspondence with density operators on $\mathcal{H}$. By contrast, the algebraic approach to QM defines states directly in terms of the observable algebra $\mathfrak{A}$, which it does not constrain to coincide with some $\mathfrak{B}(\mathcal{H})$. An algebraic *state* ω on $\mathfrak{A}$ is a linear functional $\omega: \mathfrak{A} \to \mathbb{C}$ that is normed ($\omega(I) = 1$) and positive ($\omega(A^*A) \geq 0$ for all $A \in \mathfrak{A}$). $\omega(A)$ may be understood as the expectation value of (self-adjoint) $A \in \mathfrak{A}$. Like the states of ordinary QM, the set of states on a C^*-algebra $\mathfrak{A}$ is convex. Its extremal elements—that is, states ω that cannot be expressed as nontrivial convex combinations of other states—are pure states; all other states are mixed.

Unlike ordinary quantum states, algebraic states are not required to be countably additive. In the most general case, the requirement would be otiose. The projection operators guaranteed to appear in $\mathfrak{A}$ are the additive and multiplicative identities 0 and I. With respect to that set of projection operators, countable additivity reduces to linearity and positivity.

Still, connections straightforward and surprising can be established between the Hilbert space and algebraic notions of state. Straightforwardly, a state $\widehat{W}$ acting on a Hilbert space H carrying a representation $\pi: \mathfrak{A} \to \mathfrak{B}(\mathcal{H})$ of an algebra $\mathfrak{A}$ defines an algebraic state ω on $\mathfrak{A}$. Simply, set $\omega(A) = Tr(\widehat{W} \pi(A))$ for all $A \in \mathfrak{A}$. Surprisingly, we can travel in the other direction, from an abstract algebraic state to its realization on a concrete Hilbert space. Let ω be a state on a C^*-algebra $\mathfrak{A}$. Then there exists a Hilbert space $\mathcal{H}_\omega$, a representation $\pi_\omega: \mathfrak{A} \to \mathfrak{B}(\mathcal{H}_\omega)$ of the algebra, and a cyclic[14] vector $|\Psi_\omega\rangle \in \mathcal{H}_\omega$ such that, for all $A \in \mathfrak{A}$, the expectation value the algebraic state ω assigns A is duplicated by the expectation value the Hilbert space state vector

---

14. $|\Psi\rangle$ is cyclic for $\pi_\omega(\mathfrak{A})$ means $\{\pi_\omega(\mathfrak{A}) |\Psi\rangle\}$ is *dense* in $\mathcal{H}$. That is, any vector $|\varphi\rangle \in \mathcal{H}$ can be approximated to arbitrary precision by linear combinations of elements of $\{\pi_\omega(\mathfrak{A}) |\Psi\rangle\}$.

$|\Psi_\omega\rangle$ assigns the Hilbert space observable $\pi(A)$ (in symbols, $\omega(A) = \langle \Psi_\omega|\pi_\omega(A)|\Psi_\omega\rangle$ for all $A \in \mathfrak{A}$). The triple $(\mathcal{H}_\omega, \pi_\omega, |\Psi_\omega\rangle)$ is unique up to unitary equivalence, and is the GNS (for Gelfand, Naimark, and Segal) representation of ω.

6.4. Quasi-equivalence and Disjointness

There is a sort of kinship question we might want to raise about GNS representations. Obviously, a state ω on a C^*-algebra $\mathfrak{A}$ can, by the agency of the cyclic vector $|\Psi_\omega\rangle$, be expressed in terms of ω's GNS representation. But what other states on $\mathfrak{A}$ can be expressed in terms of $(\mathcal{H}_\omega, \pi_\omega, |\Psi_\omega\rangle)$? In a clumsy and uninteresting sense, every state on $\mathfrak{A}$ can be so expressed, because every state on $\mathfrak{A}$ corresponds (via $\omega(\pi_\omega(A)) := \omega(A)$) to a normed, positive linear functional over $\pi_\omega(\mathfrak{A})$. So we had better refine our question to make it interesting. Which algebraic states on $\mathfrak{A}$ (and thus on $\pi_\omega(\mathfrak{A})$) admit natural and well-behaved extensions to the von Neumann algebra $\pi_\omega(\mathfrak{A})''$ affiliated with ω's GNS representation? It is intuitively plausible that states on $\mathfrak{A}$ (and thus on $\pi_\omega(\mathfrak{A})$) whose extensions to $\pi_\omega(\mathfrak{A})''$ are unique will be states with (roughly speaking) nice continuity properties with respect to the topologies defined on that algebra.[15] States exhibiting these continuity properties with respect to a von Neumann algebra $\mathfrak{R}$ are called *normal* states of that algebra.

A state ω on a von Neumann algebra $\mathfrak{R}$ acting on a Hilbert space $\mathcal{H}$ is normal if and only if there is a density operator $\widehat{W}$ such that $\omega(\hat{A}) = Tr(\widehat{W}\hat{A})$ for all $\hat{A} \in \mathfrak{R}$ (Bratteli and Robinson 1987, theorem [thm.] 2.4.21, 76). So states on $\mathfrak{A}$ interestingly akin to ω are states normal with respect to the von Neumann algebra affiliated with ω's GNS representation. This set of states—the set of algebraic states expressible as density matrices on ω's GNS representation—comprise what is known as a *folium*. In what follows $\mathcal{F}_\omega$ will denote ω's folium.

We will call states quasi-equivalent when their GNS representations are. If and only if states are quasi-equivalent do their folia coincide. The coincidence of the folia of quasi-equivalent states has a simple

---

15. For an elaboration, see Bratteli and Robinson 1987, 75–79.

explanation: von Neumann algebras affiliated with quasi-equivalent states are *-isomorphic; thus a state is normal on one such von Neumann algebra if and only if it is normal on the other.

Consider, by contrast, states ω and φ on $\mathfrak{A}$ whose folia have null intersection. No state on $\mathfrak{A}$ expressible as a density matrix on ω's GNS representation is so expressible on φ's GNS representation and vice versa. Such states are *disjoint*. "Disjointness," a relation that can obtain between algebraic states, radicalizes the relation of orthogonality that can obtain between vectors implementing pure Hilbert space states. Like orthogonal states, disjoint states assign one another a transition probability of 0.[16] Call states (of any stripe) *impossible relative* to one another just in case the transition probability between them is 0. Like orthogonal states, disjoint states are impossible relative to one another.

Disjointness radicalize orthogonality in the following sense. If pure states $|\varphi\rangle$ and $|\psi\rangle$ on $\mathfrak{B}(\mathcal{H})$ are orthogonal, there exist a host of pure states possible relative to both. Normed superpositions of $|\varphi\rangle$ and $|\psi\rangle$ are examples. By contrast, if ω and ρ on $\mathfrak{A}$ are disjoint, no pure state possible relative to one is possible relative to the other. This follows from the fact that pure states on $\mathfrak{A}$ are either quasi-equivalent or disjoint.

7. INTERPRETING QM_∞: SOME ANSWERS

7.1. *Hilbert Space Conservatism*

The interpreter I will label the *Hilbert space conservative* completes the kinematic template $(\mathcal{M}, \mathcal{S})$ on behalf of a theory of QM_∞ the same way it is standardly completed on behalf of a theory of ordinary QM. Our characterization of that way will be somewhat roundabout. Let $\mathfrak{A}$ be the C^*-algebra obtained by taking the norm closure of a Hilbert space representation of the CCRs/CARs for a system under consideration.

---

16. In algebraic QM, the transition probability between states ρ and ω on an observable algebra $\mathfrak{A}$ is given by $1/4 \|\omega - \rho\|^2$, where $\| \circ \|$ is the norm on the space of linear functionals on $\mathfrak{A}$ (see Roberts and Roepstorff 1969 for a discussion). This probability is 0 when ω and ρ are disjoint.

Then the Hilbert space conservative maintains that the observables pertaining to the system are the self-adjoint elements of $\pi_\omega(\mathfrak{A})''$, the von Neumann algebra affiliated with the GNS representation of a pure state ω on $\mathfrak{A}$. It follows that $\pi_\omega(\mathfrak{A})''$ coincides with $\mathfrak{B}(\mathcal{H}_\omega)$, where $\mathcal{H}_\omega$ is the concrete separable Hilbert space carrying ω's GNS representation.[17] The Hilbert space conservative identifies possible states of the system with normal states on $\pi_\omega(\mathfrak{A})''$, that is, with ω's folium $\mathcal{F}_\omega$. Because these states coincide with $\mathfrak{T}^+(\mathcal{H}_\omega)$, the recapitulation of ordinary QM is complete. When the Hilbert space conservative offers $(\pi_\omega(\mathfrak{A})'', \mathcal{F}_\omega)$ as the kinematic pair for a theory of QM_∞, she or he is just dressing the kinematic pair $(\mathfrak{B}(\mathcal{H}), \mathfrak{T}^+(\mathcal{H}))$ typical of ordinary QM up in algebraic garb.

The Hilbert space conservative also transfers ordinary QM's content coincidence concept of physical equivalence to the setting of QM_∞. Theories specified up to kinematic pairs $(\pi_\omega(\mathfrak{A})'', \mathcal{F}_\omega)$ and $(\pi_{\omega'}(\mathfrak{A})'', \mathcal{F}_{\omega'})$ are physically equivalent, the Hilbert space conservative maintains, just in case $(\pi_\omega(\mathfrak{A})'', \mathcal{H}_\omega)$ and $(\pi_{\omega'}(\mathfrak{A})'', \mathcal{H}_{\omega'})$ are unitarily equivalent, from which it follows that $\mathcal{F}_\omega = \mathcal{F}_{\omega'}$.

Of course having said this much, the Hilbert space conservative is hardly done interpreting QM_∞. In ordinary QM, the uniqueness theorems assure us, once $\mathfrak{A}$ is fixed, all representations are unitarily equivalent. This is not so in QM_∞. The algebra $\mathfrak{A}$ that fixes a system's type admits continuously many unitarily inequivalent Hilbert space representations, and there are as many candidate completions of the kinematic template $(\pi_\omega(\mathfrak{A})'', \mathcal{F}_\omega)$ as there are disjoint pure states ω on $\mathfrak{A}$. Committed to unitary equivalence as a criterion of physical equivalence, the Hilbert space conservative must regard each of these as rival quantum theories. At most one of them can be the theory she or he has set out to interpret. Thus the burden of the Hilbert space conservative is to articulate and motivate a principle of privilege that enables her or him to enthrone a single unitary equivalence class of representations as physical and to consign the rest to the dustbin of mere mathematical artifacts.

---

17. This follows via considerations so far suppressed: the GNS representation of a pure state ω is irreducible (Bratteli and Robinson 1987, thm. 2.3.19, 57)—that is, the only subspaces of $\mathcal{H}_\omega$ invariant under π_ω are the zero subspace and $\mathcal{H}_\omega$ itself—and $\pi_\omega(\mathfrak{A})''$ is isomorphic to $\mathfrak{B}(\mathcal{H}_\omega)$ if π_ω is irreducible (Sakai 1971, thm. 1.12.9, 19).

Promising principles of privilege apply to quantum field theories and appeal to the symmetries of the spacetimes in which they are set. In a spacetime equipped with global timelike isometries (i.e., symmetries of the metric), a unitary equivalence class of representations of CCRs for the Klein-Gordon field exhibits energetics respecting those isometries. The respect is expressed by representations with the particle-interpretation-tempting structure sketched in section 4 and takes roughly[18] the following form: a privileged Hamiltonian operator can be identified as the infinitesimal generator of time translations that leave the metric unchanged. A respectful representation has a vacuum state that is not only itself invariant under these timelike isometries but also the lowest-energy eigenstate of the privileged Hamiltonian. What is more, each n-particle state of a respectful representation is an eigenstate of the privileged Hamiltonian, with an eigenvalue matching the energy of a corresponding classical state. Bernard S. Kay and Robert M. Wald (1991) show that in suitably symmetrical spacetimes, respectful representations of the Klein-Gordon CCRs are unique up to unitary equivalence and have a particle-interpretation-tempting structure.

If it is reasonable to demand such respect, it is reasonable to distinguish invidiously between respectful and disrespectful representations. The invidious distinction makes available univocal and powerful particle talk. Particle notions incommensurable with, because circumscribed by representations unitarily inequivalent to, that of the respectful representation can be dismissed as unphysical. Each state in the privileged folium can be characterized in terms of its particle content; different states can be distinguished in terms of their particle content. If an adequate interpretation is obliged to understand QFT as particle physics—and that is how many of its practitioners do understand it—then the pursuit of Hilbert space conservatism in spacetime settings with enough symmetry to pick out a particle-friendly unitary equivalence class of representations is one route to adequacy.

But it is a route a that is blocked for QFTs set in spacetimes lacking the relevant symmetries and a route that bypasses the thermodynamic

---

18. See Kay and Wald 1991 for a rigorous exposition.

limit of QSM, which is set in Euclidean three space. There, as we will see presently, Hilbert space conservatism is untenable.

Short of the thermodynamic limit and in the setting of concrete Hilbert spaces, the *Gibbs state* equips QSM with a notion of equilibrium. The Gibbs state describing a system governed by Hamiltonian $\hat{H}$ in equilibrium at inverse temperature $\beta = 1/kT$ is

$$\hat{\rho} = \frac{\exp(-\beta \hat{H})}{Tr[\exp(-\beta \hat{H})]} \quad (15)$$

For realistic, finite quantum systems this Gibbs state is well-defined and unique (Ruelle 1969). For infinite quantum systems this need not be so: the r.h.s. of (15) fails to be well defined.

Suppose that we aspire to construct a quantum statistical account of *phase structure*. Then we would have reason to concoct a notion of equilibrium suited to infinite systems. For the apparent macroscopic explanandum is the existence, at certain temperatures, of multiple thermodynamic phases. If a statistical account of phase structure requires the existence, at these critical temperatures, of *multiple distinct* equilibrium states answering to different thermodynamic phases, then, as long as we are dealing with finite systems, the very uniqueness of the Gibbs state frustrates our explanatory aspirations.

Explanatory hopes are revived in the thermodynamic limit by using the *KMS* (for Kubo, Martin, and Schwinger) *condition* to explicate a notion of equilibrium more general than that afforded by the Gibbs state.[19] The set of KMS equilibrium states with respect to a given dynamics has several striking features.[20] One is that a system of type $\mathfrak{A}$ in equilibrium at a temperature T occupies a state *disjoint* from the state of the same system in equilibrium at a different temperature (Bratteli and Robinson 1997, thm. 5.3.35, 128). Provided neither temperature is 0, each equilibrium state is moreover—and mind-bogglingly—disjoint from every

---

19. For more on the KMS condition and why it suitably explicates equilibrium, see Bratteli and Robinson 1997, sec. 5.3.
20. The following statements suppress certain technical assumptions that generally hold at the thermodynamic limit; see the references for details.

pure state on $\mathfrak{A}$.[21] And at critical temperatures where $\mathfrak{A}$ admits *multiple* KMS states, the extremal elements of that set—which correspond to the pure phases whose copresence constitutes the phase structure we aspire to explain (cf. Sewell 1984, sec. 4.4)—are either equivalent or disjoint (Bratteli and Robinson 1997, thm. 5.3.30, 116–117).

Holding that physically possible states of a system of type $\mathfrak{A}$ lie in the folium of some pure state ω on $\mathfrak{A}$, Hilbert space conservatism paralyses the thermodynamic limit of QSM. Hilbert space conservatism implies that equilibrium at temperatures different from 0 is impossible, because no such equilibrium state lies in the folium of a pure state. Even if Hilbert space conservatism is relaxed to allow an *impure* state φ to pick out the privileged folium, it implies that at most *one* equilibrium state at a nonzero temperature—the one given by φ—is possible. This declaration hamstrings the attempt to explain phase structure in terms of the availability, at the critical temperatures at which they occur, of multiple distinct equilibrium states.

Even this preliminary review of Hilbert space conservatism suggests that its viability depends on *which* theory of QM_∞—QFT in Minkowski spacetime? QFT in a less symmetrical spacetime? The thermodynamic limit of QSM?—one is interpreting. This sensitivity of viable interpretive options to particularities, even peculiarities, of the theory interpreted will be an ongoing theme.

7.2. *Algebraic Imperialism*

According to an interpretive strategy Aristidis Arageorgis (1995) has dubbed "algebraic imperialism," the extra structure one obtains along with a concrete representation of (for example) the Weyl algebra is extraneous—the Weyl algebra, and not any of its concrete but disparate Hilbert space realizations, encapsulates the physical content of QFT. For Irving E. Segal (1967, 128), a pioneer of C^*-algebraic approaches, "The important thing here is that the observables form some algebra, and not the representation Hilbert space on which they act." Inspired by the representation independence of the Weyl algebra, algebraic imperialism construes a theory of QM_∞ not in terms of the Hilbert space

---

21. A circumstance discussed in Ruetsche 2004.

setting of a particular concrete representation of the CARs/CCRs for that theory but in terms of the abstract algebraic structure every such representation shares. Physical magnitudes pertaining to a system of type $\mathfrak{A}$ are given by the self-adjoint part of the abstract C^*-algebra $\mathfrak{A}$; possible states are states in the algebraic sense on $\mathfrak{A}$, which I will denote $S_\mathfrak{A}$. ($\mathfrak{A}, S_\mathfrak{A}$) thus gives the kinematic pair for a theory of QM_∞.

Algebraic imperialism and Hilbert space conservatism are rival accounts of the content of QM_∞. For there can arise algebraic states ω and ω' on an observable algebra $\mathfrak{A}$ of QM_∞ that are disjoint. Both are, according to the algebraic imperialist, possible states of the system associated with $\mathfrak{A}$. But they cannot both lie in the folium of some pure state ω on $\mathfrak{A}$. If unitary equivalence is criterial for physical equivalence, any theory admitting one state as possible will be physically inequivalent to a theory admitting the other as possible.

7.3. Apologetic Imperialism

This makes it clear that unitary equivalence is a criterion of physical equivalence ill adapted to algebraic imperialism. Unitary equivalence distinguishes invidiously between disjoint folia; the algebraic approach does not. Recognizing a mismatch, early advocates of the algebraic approach (see, e.g., Haag and Kastler 1964, 851; cf. Robinson 1966, 488) offered instead of unitary equivalence *weak equivalence* as a criterion of physical equivalence for Hilbert space representations. It seems to me that these advocates regarded their recognition of unitarily inequivalent representations as requiring some sort of apology, an apology they couched in terms of weak equivalence. So I will call them *apologetic imperialists*.

Weak equivalence is a relation that holds (or not) between Hilbert space representations of a C^*-algebra $\mathfrak{A}$. Two representations are weakly equivalent exactly when no finite set of expectation values specified to finite accuracy can locate a state in one, rather than the other, representation. More precisely, $(\mathcal{H}, \pi)$ and $(\mathcal{H}', \pi')$ are weakly equivalent if and only if, given any Hilbert space state on the first representation and the algebraic state ω it yields and given any finite set of elements of the abstract algebra of observables and any finite set of nonzero margins of experimental error, there exists a Hilbert space state on the second representation yielding an algebraic state ω' that reproduces within those margins of errors ω's assignment of expectation values to those observables.

Now the apologizing imperialist "define[s] two representations to be *physically equivalent* if and only if they are weakly equivalent" (Robinson 1966, 488). A rough and ready operationism suggests a justification of the equation of physical and weak equivalence. "We want the results of any finite set of measurements on a physical state to be equally well describable in terms of a density matrix on $\mathcal{H}_1$ or a density matrix on $\mathcal{H}_2$. As measurements, are never totally accurate, 'equally well' is to be understood as 'to any desired degree of accuracy'" (Kastler 1964, 180–181).

Any actual measurement performed on a system in the algebraic state ω will fix the expectation values of some finite collection of observables A_i, $i = 1, 2,\ldots, n$ within experimental error margins $\varepsilon_i > 0$. Accordingly, such a measurement cannot distinguish ω from any other state ω' satisfying $|\omega(A_i) - \omega'(A_i)| < \varepsilon_i$ for all i. Such states are, the apologetic imperialist contends, for all practical purposes, physically equivalent. And so weakly equivalent representations are for all practical purposes physically equivalent as well.

A mathematical result known as Fell's theorem consolidates weak equivalence as a criterion of physical equivalence hospitable to algebraic approaches. Fell's theorem implies that all faithful representations of a C^*-algebra are weakly equivalent (see Wald 1994, chap. 4.5). Defining physical equivalence as weak equivalence, the apologetic imperialist can conclude that "all faithful representations of [the Weyl algebra] are physically equivalent" (Robinson 1966, 488)—and chastise the conservative for elevating surplus structure to physical significance: "It is in this new notion of equivalence in field theory that the algebraic approach has its greatest justification. All the physical content of the theory is contained in the algebra itself; nothing of fundamental significance is added to a theory by its expression in a particular representation."

Notice that the observables with respect to which weakly equivalent representations are practically indistinguishable are elements of the abstract algebra $\mathfrak{A}$. Consider, for some concrete representation π_1 of $\mathfrak{A}$, an observable in the weak closure of that representation but without correlate in $\mathfrak{A}$. I will call such an observable *parochial* to the representation π_1. Examples of parochial observables include the total number operators (or more properly the projectors in their spectral resolution) encountered in suitable representations of the Weyl algebra for the free Klein-Gordon field. There is a π_1-normal state, the vacuum state, which assigns the total number operator expectation value 0. But there is no normal state in any representation π_2 unitarily inequivalent to π_1 that assigns this total number operator an expectation value within any finite ϵ of 0 (Clifton and Halvorson 2001, prop. 11, 450). Every faithful representation of the Weyl algebra is weakly equivalent to every other. So apologetic imperialists would hail π_1 and π_2 as physically equivalent, even though all π_2-normal states assign an observable in the weak closure of π_1—its total number operator—"expectation values"[22] arbitrarily *far* from that assigned by the π_1 vacuum state. Of course this is a compelling criticism of weak equivalence as an explication of physical equivalence only for those inclined—as imperialists are *not*—to regard parochial observables as physical.

The justifications apologetic imperialists offer for weak equivalence as a criterion of physical equivalence for Hilbert space representations are, broadly speaking, operationalist. But they ultimately fail, *even by operationalist lights*. For, as Stephen J. Summers (2001) has observed, even operationalists ought to take states to be predictive instrumentalities. Operationalists should therefore expect the equivalence of states to extend to their predictions concerning future measurements. But Fell's theorem offers us no assurance that a π_1-normal state for all practical purposes indistinguishable from a π_2-normal state with respect to some set $\{A_i\}$ of algebraic observables will continue to mimic the first state's predictions with respect to an expanded or altered set of observables. The practical indistinguishability of states that secures the weak equivalence of the representations bearing them is only a backward-looking or static sort of indistinguishability. Before recognizing a normal state on one representation as a suitable counterpart to a normal state on another, an operationalist should demand not only backward-looking

---

22. The scare quotes because, strictly speaking, the observable is outside the domain of the state.

but forward-looking indistinguishability: the indistinguishability of putative counterpart states conceived as predictive instrumentalities. Thus the appeal to Fell's theorem should not convince even the operationalist that weak equivalence is an acceptable notion of physical equivalence.

7.4. Bold Imperialism

Apologies for algebraic liberality invoking weak equivalence fall flat. Insofar as weak equivalence is a criterion of equivalence for Hilbert space representations—which the imperialist *denies* to be repositories of physical content—these apologies were heading in the wrong direction to begin with. A bolder imperialism would define physical equivalence directly in terms of the imperialist kinematic pair $(\mathfrak{A}, S_\mathfrak{A})$. Two such pairs $(\mathfrak{A}, S_\mathfrak{A})$ and $(\mathfrak{A}', S_{\mathfrak{A}'})$ are equivalent in the content coincidence sense (explicated for ordinary QM by [9]) only if their associated algebras are isomorphic. Once $(\mathfrak{A}, S_\mathfrak{A})$ is identified as the appropriate kinematic pair for a theory of QM_∞, this criterion of physical equivalence follows straightforwardly. The challenge for the bold imperialist is to justify her or his choice of kinematic pair.

Segal, possibly the boldest imperialist of them all, rises to the challenge by an operationalist ploy of his own. All hands, Segal suggests, should grant that the canonical observables pertaining to a system are physically meaningful and grant physical meaning as well to bounded functions of finite sets of canonical observables. Segal's crucial move is to grant physical significance to the norm closure but not (where they differ) the weak closure of this set of antecedently meaningful observables. He accepts that "[norm] approximation is operationally meaningful," because where f_n norm approximates f, f's "expectation value in any state is simply the limit of the expectation values of the approximating bounded functions" (Segal 1959, 347–348). To see the point of this observation, imagine a sequence of A_i ∈ $\mathfrak{A}$ that does not converge in $\mathfrak{A}$'s norm topology. Suppose further that in some representation π of $\mathfrak{A}$, the sequence $\pi(A_i)$ *does* weakly converge to an operator A. Recall that π-normal states on $\mathfrak{A}$ are those states φ whose extensions from $\pi(\mathfrak{A})$ to its weak closure $\pi(\mathfrak{A})''$ are unique. So a *π-abnormal* state ω on $\mathfrak{A}$ can be such that the sequence $\omega(A_i)$ fails to converge even though the sequence $\pi(A_i)$ converges weakly. Segal contends that A, and other parochial observables, lack operational significance, because there are states whose expectation value assignments to observables conceded significance by all hands do not fix their expectation value assignments to A. Segal deems norm limits admissible, because if A_i norm converges to some limit A, $\omega(A_i)$ perforce converges as well.

The keystone difference, for Segal, between weak and norm limits, then, is that the world can get itself into a condition—the condition of a state not normal with respect to the representation whose weak topology is used to take a weak limit—where the expectation value of a weak limit floats free of any set, even an infinite one, of laboratory machinations, even perfectly accurate ones. With norm limits, this is not so. Thus Segal would vest norm but not weak limits with operational and so physical significance. It follows that he regards as significant the C^*, but not (where they differ) the von Neumann, algebra generated by elements satisfying the canonical relations defining a system and regards as equivalent kinematic templates plying isomorphic C^*-algebras. Therein lies his bold imperialism.

Segal's derivation of imperialism from operationalism impresses me as deeper and less opportunistic than the apologetic imperialists' plea for weak equivalence. Segal rests his case on a principle like:

(SEGOP) Let $\mathcal{P}$ be the set of bounded functions of canonical observables for a system S. X is an observable pertaining to S if and only if there is a sequence $X_i \in \mathcal{P}$ such that, for any state ω of S, $\omega(X_i)$ converges. When this condition is met, $\lim_{i \to \infty} \omega(X_i)$ is the expectation value of X in the state ω.

Reflecting the conviction that physical magnitudes are those for which there exist measurement protocols, SEGOP has operationalist roots. But the principle makes minimal appeal to the depressing actualities of laboratory life—its finitude, imperfect material conditions, or inaccuracy. Indeed a confirmed metaphysician who forswears operationalism might nevertheless accept (SEGOP) as an explication of the parsimony desideratum discussed near the close of section 5. (SEGOP) is telling us which magnitudes we are committed to, once we have adopted a set of canonical magnitudes and space of possible states, and which magnitude pretenders turn out to be free-riding fluff.

Of course the catch is that to apply Segal's criterion we need to know which states are possible. It follows that (SEGOP) will not help us settle the sort of interpretive dispute that consists in disagreeing, as the Hilbert space conservative and the algebraic imperialist disagree, about which states are possible to begin with. A Hilbert space conservative, confining possible states for a system of canonical type $\mathfrak{A}$ to those in the folium of a pure state ω on $\mathfrak{A}$ can appeal to SEGOP to earn observables parochial to $\pi_\omega(\mathfrak{A})$ physical significance. For she or he recognizes no π_ω-abnormal states that might interact with SEGOP to disqualify those observables.

SEGOP sets a tenet of interpretive good taste (if not a test of internal coherence) for an interpretation. But it cannot adjudicate between competing interpretations that meet its scruples. We will have to fall back on our vaguer criteria to call those contests.

7.5. Mixed Strategies

By a "mixed" strategy for interpreting QM_ω, I mean one that mixes and matches elements of the kinematic pairs for the imperialist and conservative approaches. Two mixed strategies are discussed in Clifton and Halvorson 2001, 430ff.. One offers $(\mathfrak{A}, \mathcal{F}_\omega)$ as the kinematic template for a system of canonical type $\mathfrak{A}$. The other offers $(\pi_\omega(\mathfrak{A})'', S_\mathfrak{A})$. Let us focus on the latter, which concedes to the conservative that observables reside in the von Neumann algebra generated by some concrete representation but suspends the conservative's requirement that states be countably additive.

Our discussion of the Hilbert space conservative foreshadows one challenge for this mixed strategy. Suppose π_ω is the representation whose affiliated von Neumann algebra contains the physical magnitudes. If ω and ω' are disjoint, some self-adjoint elements of the von Neumann algebra affiliated with $\pi_{\omega'}$ are disqualified from physical significance. The challenge is to justify this disqualification. In a way the challenge is deeper than the Hilbert space conservative's problem of privilege. The conservative can at least observe a SEGOP-style consonance between the states she or he recognizes and the observables she or he recognizes. The mixed strategy enjoys no such consonance. In $S_\mathfrak{A}$ are states whose values on $\pi_\omega(\mathfrak{A})$ fix their values on $\pi_\omega(\mathfrak{A})''$ and states for which this is not so. So it seems our discussion of bold algebraic imperialism foreshadows a problem for the mixed strategy as well: it stands in violation of SEGOP. In a sense Segal has made precise, the strategy recognizes too many states for her or his observable set to bear.

7.6. Universalism

The last section found mixed strategies wanting not only in motivation but also in the sort of coherence articulated by SEGOP. This section sketches an interpretive stance I call "universalism." Universalism construes the content of a quantum theory of type $\mathfrak{A}$ by means of the extravagance of the *universal representation* $\mathfrak{U}$ of $\mathfrak{A}$. The universal representation is the direct sum,[23] over states on $\mathfrak{A}$, of their GNS representations:

$$\pi_U = \oplus_{\omega \in S_\mathfrak{A}} \pi_\omega \qquad (16)$$

π_U acts on the direct sum Hilbert space $\mathcal{H}_u = \oplus_{\omega \in S_\mathfrak{A}} \mathcal{H}_\omega$. The *universal enveloping von Neumann algebra* $\mathfrak{U}$ is the weak closure of $\pi_U(\mathfrak{A})$. Universalism offers $\mathfrak{U}$ as the algebra of observables pertaining to a system of type $\mathfrak{A}$ and normal states on $\mathfrak{U}$ as possible states of such a system.[24] $\mathfrak{U}$ has the nice feature that its normal states coincide with $S_\mathfrak{A}$ (Emch 1972, 120–121).

---

23. Where $\mathcal{H}$ and $\mathcal{H}'$ are Hilbert spaces, their direct sum $\mathcal{H} \oplus \mathcal{H}'$ consists of all elements of the form $|a\rangle \oplus |\beta\rangle$, $|a\rangle \in \mathcal{H}$ and $|\beta\rangle \in \mathcal{H}'$. The rule for vector addition is $|a_1\rangle \oplus |\beta_1\rangle + |a_2\rangle \oplus |\beta_2\rangle = (|a_1\rangle + |a_2\rangle) \oplus (|\beta_1\rangle + |\beta_2\rangle)$. Inner products work like so: $(\langle a_1| \oplus \langle \beta_1|)(|a_2\rangle \oplus |\beta_1\rangle + |\beta_2\rangle) = \langle a_1|a_2\rangle + \langle \beta_1|\beta_2\rangle$. $\mathcal{H} \oplus \mathcal{H}'$ contains *only* those elements: the direct sum of two Hilbert spaces corresponds to their Cartesian, rather than their tensor, product. See Emch 1972, 120ff. for more on the universal representation.
24. See Kronz and Lupher 2005 for one version of this proposal.

The conservative commits invidious privilege. The universalist (like her or his religious counterpart) is adamantly noncommittal. The imperialist haughtily ignores representation-specific observables that could be wage-earning members of physical theories. The universalist welcomes any observable that can make it into the weak closure of the universal representation.[25] Mixed strategies run afoul of SEGOP, a principle of interpretive good taste universalism promises to obey. Notwithstanding this litany of accomplishments, Universalism has struck many as excessive (cf. Kastler 1964, 184; Takesaki 2000, 157). π_U represents each element A of $\mathfrak{A}$ continuously many times: once as $\pi_U(A)$ and once for each proper subrepresentation of π_U. Each representation duplicates every other's functional relationships to canonical observables. Which is the real A? The answer $\oplus_\omega \pi_\omega(A)$, in effect a disjunction of individually parsimonious options, will not satisfy anyone inclined to ask the question to begin with.

7.7. Tempered Universalism

For a system of type $\mathfrak{A}$, the approach I call "tempered universalism" advances some criterion of privilege that identifies as physically reasonable a subset Φ of the full set $S_\mathfrak{A}$ of algebraic states. The tempered universalist then constructs an algebra of physical magnitudes as the direct sum of the von Neumann algebras affiliated with states in the privileged set. The hope is that, with the appropriate principle of tempering, this strategy will prove less extravagant than universalism but more generous than either conservatism or the imperialism.

Candidate tempering principles include:

7.7.1. SYMMETRY MONGERING

In the presence of spacetime symmetries Λ implemented by automorphisms a_λ of $\mathfrak{A}$, a state ω on $\mathfrak{A}$ is physical only if each symmetry a_λ can be

---

25. NB this may be fewer observables than you imagine. It follows from theorem 10.3.5 of Kadison and Ringrose 1986, 738 that $\pi_U(\mathfrak{A})'' \neq \oplus_{\phi \in S} \pi_\phi(\mathfrak{A})''$. So "portmanteau" operators like $\pi_\phi(X) \oplus_{\psi \neq \phi} 0$—the sort of operators we would expect to implement π_ϕ-parochial observables on $\mathcal{H}_\omega$—will not necessarily make it into $\mathfrak{U}$.

implemented unitarily in π_ω. α can be implemented unitarily in $(\pi, \mathcal{H})$ if and only if there exists a unitary $U \in \mathfrak{B}(\mathcal{H})$ such that

$$\pi(\alpha(A)) = U_\pi A U^{-1} \text{ for all } A \in \mathfrak{A} \qquad (17)$$

For an early instance, D. W. Robinson (1966, 483) proposes a principle of this sort. Our discussion of promising criteria of privilege for the Hilbert space conservative shows how symmetry mongering might pay off. When a_t is a one-parameter family of symmetry-implementing automorphisms of $\mathfrak{A}$, the family of unitaries U_t demanded by the principle could have as an infinitesimal generator a magnitude that does physical work.

7.7.2. ACTUALISM

A state ω on $\mathfrak{A}$ is physical only if it is the actual state of the system of interest or appropriately related to that state.

Perhaps Hans Primas (1983, 174–175) suggests something of this sort when he writes: "To select a particular representation means to concentrate the attention to a particular phenomenon and to forget effects of secondary importance. With such a choice (which is possible if we adopt an appropriate topology in our mathematical formalism) we lose irrelevant information so that the system becomes simpler." This principle of tempering threatens to fall victim to the complaint voiced about Hilbert space conservatism that it excludes too many possible states. One version of that complaint pivoted on the fact that equilibrium states at different temperatures are disjoint. Universalism tempered by an actualism that failed to exploit the "appropriately related" clause would commit the modal solecism of declaring equilibria at temperatures other than the actual one impossible.

However, exploiting the appropriately related clause is a delicate business. A natural suggestion is that ω and ω' are appropriately related if they are dynamically accessible one from the other, that is, only if there is a "dynamical automorphism" a_t of $\mathfrak{A}$ s.t. $\omega = \omega' \circ a_t$. But which automorphisms are dynamical? Given that the algebraic approach countenances nonunitarily implementable dynamics (documented in

Arageorgis et al. 2002), one fraught question is whether these automorphisms must be unitarily implementable on the GNS representations of the states in question to qualify.

7.7.3. HADAMARDISM

A state ω on $\mathfrak{A}$ is physical only if it satisfies the Hadamard condition, which means it supports a procedure for assigning an expectation value to a stress-energy observable. Robert M. Wald (1994, sec. 4.6) announces a principle of this sort. For reasons outlined in section 5, this tempering principle makes the most sense for quantum field theories consorting with semiclassical quantum gravity. But there it can have dramatic consequences. In spacetimes with compact Cauchy surfaces (or closed universes), all Hadamard vacuum states are unitarily equivalent (Wald 1994, 97). In open universes, this is not so.

7.7.4. DYNAMICISM

A state ω on $\mathfrak{A}$ is physical only if it supports dynamics. The cases to which this principle applies most directly come from the thermodynamic limit of quantum statistical mechanics. There a dynamics for subalgebras $\mathfrak{A}_V$ of the overall algebra $\mathfrak{A}$—subalgebras associated with finite volumes V—may not have a *norm* limit as $V \to \infty$ but may have a weak limit in certain representations π_ω of $\mathfrak{A}$ (see Sewell 2002, sec. 2.4.5). The idea behind this principle is that states whose GNS representations enable a well-defined global dynamics are more physical than those that frustrate such dynamics.

7.7.5. DHR SELECTION THEORY

DHR (Doplicher, Haag, Roberts) selection theory is an approach to superselection in algebraic QFT that identifies the physical representations of the field algebra $\mathfrak{A}$ with those whose observable differences from a fixed vacuum representation π_0 are confined to local regions. Considering the set of DHR representations from the vantage point of category theory, one can then *reconstruct* the quantum fields and the gauge group. For more about this powerful and profound approach, see Halvorson and Müger 2007. Abstracting away from the details, observe that this principle of selection proceeds from a preidentified vacuum state ω_0.

8. CONCLUSION: TEMPERING AND TOGGLING

What all these tempering strategies have in common is that they complicate the modal toggle picture of a physical theory. On that picture, a physical theory sorts logically possible worlds into those that are (according to its kinematics) physically possible and those that are not. These physically possible worlds are, as it were, instantaneous; the theory's dynamical laws tell us which physically possible worlds are the time developments of which others. If the theory is lucky enough to be true, uninteresting contingency tells us which trajectory the actual world lies on. Different untempered interpretations offer different sorting mechanisms: a Hilbert space structure of observables, an abstract algebraic structure, a universal representation structure, and so on. But they deploy these sorting mechanisms, as it were, a priori. Conservatism, imperialism, mixed strategies, and universalism did not need to peek at a theory's spacetime setting, its symmetries, its dynamics, or its relation to semiclassical quantum gravity to complete the kinematic template on its behalf. They aspire to a greater level of generality than that.

Tempered universalism does not. Tempered universalism peeks. Different tempering principles peek in different places, but they all peek. Peeking, each gives up on the idea that what is kinematically possible according to a quantum theory has everything to do with that theory's kinematical laws (that is, [anti]commutation relations) and nothing to do with its nonkinematic particulars. I take tempered universalism to suggest that to apply a theory of QM_∞ to a particular problem is not simply to select appropriate elements of a preconfigured set of worlds possible according to that theory. Rather, the configuration of that set can happen along with the application of the theory and happen in different ways for different applications. The toggle is a myth that obscures the resourcefulness of theories of QM_∞.

References

Albert, David. 1994. *Quantum Mechanics and Experience*. Cambridge, MA: Harvard University Press.

Arageorgis, Aristitidis. 1995. "Fields, Particles, and Curvature: Foundations and Philosophical Aspects of Quantum Field Theory in Curved Spacetime." PhD diss., University of Pittsburgh.

Arageorgis, Aristidis. John Earman, and Laura Ruetsche. 2002. "Weyling the Time Away: The Non-Unitary Implementability of Quantum Field Dynamics on Curved Spacetime." *Studies in History and Philosophy of Modern Physics* 33:151–184.

Arageorgis, Aristidis, John Earman, and Laura Ruetsche. 2003. "Fulling Non-uniqueness and the Unruh Effect: A Primer on Some Aspects of Quantum Field Theory." *Philosophy of Science* 70:164–202.

Bratteli, Ola, and Derek W. Robinson. 1987. *Operator Algebras and Quantum Statistical Mechanics*. Vol. 1. 2nd ed. Berlin: Springer-Verlag.

Bratteli, Ola, and Derek W. Robinson. 1997. *Operator Algebras and Quantum Statistical Mechanics*. Vol. 2. 2nd ed. Berlin: Springer-Verlag.

Brighouse, Carolyn. 1994. "Spacetime and Holes." In *PSA: Proceedings of the Biennial Meeting of the Philosophy of Science Association*. Vol. 1, *Contributed Papers*, ed. David M. Hull, Mickey Forbes, and Richard Burian, 117–125. East Lansing, MI: Philosophy of Science Association.

Clifton, Robert, and Hans Halvorson. 2001. "Are Rindler Quanta Real? Inequivalent Particle Concepts in Quantum Field Theory." *British Journal for the Philosophy of Science* 52:417–470.

Emch, Gérard G. 1972. *Algebraic Methods in Statistical Mechanics and Quantum Field Theory*. New York: Wiley.

Haag, Rudolf, and Daniel Kastler. 1964. "An Algebraic Approach to Quantum Field Theory." *Journal of Mathematical Physics* 5:848–861.

Halvorson, H. 2004. "Complementarity of Representations in Quantum Mechanics." *Studies in History and Philosophy of Modern Physics* 35:45–56.

Halvorson, Hans, and Michael Müger. 2007. "Algebraic Quantum Field Theory." In *Philosophy of Physics*, ed. Jeremy Butterfield and John Earman, 731–922. Amsterdam: Elsevier.

Hempel, Carl. 1965. "The Theoretician's Dilemma. In *Aspects of Scientific Explanation and Other Essays in the Philosophy of Science*, 173–226. New York: Free Press.

Huggett, Nick. 2000. "Philosophical Foundations of Quantum Field Theory." *British Journal for the Philosophy of Science* 51:617–637.

Kadison, Richard V., and John R. Ringrose. 1986. *Fundamentals of the Theory of Operator Algebras*. Vol. 2. New York: Academic Press.

Kastler, Daniel. 1964 "A C*-algebra Approach to Field Theory." In *Proceedings of a Conference on the Theory and Application of Analysis in Function Space Held at Endicott House in Dedham, Massachusetts, June 9–13, 1963*, ed. William Ted Martin and Irving E. Segal, 179–191. Cambridge, MA: MIT Press.

Kay, Bernard S. and Robert M. Wald. 1991. "Theorems on the Uniqueness and Thermal Properties of Stationary, Nonsingular, Quasifree States on Spacetimes with a Bifurcate Killing Horizon." *Physics Reports* 207:49–136.

Kronz, Frederick M., and Tracy A. Lupher. 2005. "Unitarily Inequivalent Representations in Algebraic Quantum Theory." *International Journal of Theoretical Physics* 44:1239–1258.

Marsden, Jerrold E., and Tudor S. Raițu. 1994. *Introduction to Mechanics and Symmetry*. New York: Springer-Verlag.

Primas, Hans. 1983. *Chemistry, Quantum Mechanics, and Reductionism*. New York: Springer-Verlag.

Roberts, J. E., and G. Roepstorff. 1969. "Some Basic Concepts of Algebraic Quantum Theory." *Communications in Mathematical Physics* 11:321–338.

Robinson, D. W. 1966. "Algebraic Aspects of Relativistic Quantum Field Theory." In *Axiomatic Field Theory*, ed. M. Chretien and S. Deser, 389–516. New York: Gordon and Breach.

Ruelle, David. 1969. *Statistical Mechanics*. New York: Benjamin.

Ruetsche, Laura. 2004. "Intrinsically Mixed States: An Appreciation." *Studies in History and Philosophy of Modern Physics* 35:221–239.

Ruetsche, Laura. 2011. *Interpreting Quantum Theories*. Oxford: Oxford University Press.

Sakai, Shôichirô. 1971. *C*-algebras and W*-algebras*. New York: Springer-Verlag.

Segal, Irving E. 1959. "The Mathematical Meaning of Operationalism in Quantum Mechanics." In *Studies in Logic and the Foundations of Mathematics*, ed. L. Henkin, P. Suppes, and A. Tarski, 341–352. Amsterdam : North-Holland.

Segal, Irving E. 1967. "Representations of the Canonical Commutation Relations." In *Application of Mathematics to Problems in Theoretical Physics*, ed. François Lurçat, 107–170. Cargèse Lectures in Theoretical Physics. New York: Gordon and Breach.

Sewell, Geoffrey L. 1984. *Quantum Theory of Collective Phenomena*. New York: Oxford University Press.

Sewell, Geoffrey L. 2002. *Quantum Mechanics and Its Emergent Metaphysics*. Princeton, NJ: Princeton University Press.

Sklar, Lawrence. 2002. *Theory and Truth*. Oxford: Oxford University Press.

Summers, Stephen J. 2001. "On the Stone–von Neumann Uniqueness Theorem and Its Ramifications." In *John von Neumann and the Foundations of Quantum Physics*, ed. Miklós Rédei and Michael Stöltzner, 135–152. Vienna Circle Institute Yearbook 8. Dordrecht, Netherlands: Kluwer Academic.

Takesaki, Masamichi. 2000. *Theory of Operator Algebras I*. Berlin: Springer Verlag.

Teller, Paul. 1995. *An Interpretive Introduction to Quantum Field Theory*. Princeton, NJ: Princeton University Press.

Wald, Robert M. 1994. *Quantum Field Theory in Curved Spacetime and Black Hole Thermodynamics*. Chicago: University of Chicago Press.

Chapter 9

Statistical Mechanics in Physical Theory

LAWRENCE SKLAR

1. THERMODYNAMIC CONCEPTS AND EXPLANATIONS

Thermodynamics begins with the empirical study of temperature as turned into a quantifiable matter by the invention of the thermometer and with the phenomenological ideal gas laws. The subject becomes far richer when the interconversion of stored heat with mechanical work, so essential for the practical purpose of the running of a steam engine, becomes a theoretical matter at the hands of Sadie Carnot.

A long process shows that there is a way of generalizing the conservation of mechanical energy familiar from mechanics into a broader principle in which internal, hidden energy is attributed to a substance and the sum of this hidden energy with overt mechanical energy is posited as conserved in the first law of thermodynamics. With the efficiency of an engine limited by the difference in temperature of heat source and heat sink, the temperature scale becomes one founded on an absolute zero as the temperature of sink in which ideally all the heat available in the source can be converted into usable work.

The deep insight is obtained that the total amount of heat energy is not the sole factor that matters. There is its availability for doing mechanical work, thought of as its existence in an "uneven" condition. So the heat in the high temperature reservoir can do work if a lower temperature condenser is at hand, but uniformly distributed heat is useless for mechanical work. This becomes elegantly formulated by Rudolph Clausius's concept of entropy (with low entropy signifying availability for work) and with the famous second law. Here the most profound realization is that thermally isolated systems left to themselves can spontaneously move to higher-entropy, more uniform conditions. But a spontaneous movement to a lower-entropy condition of an isolated system is not seen. The end state toward which systems tend spontaneously is an unchanging equilibrium state. This time asymmetry of a fundamental physical law is a striking novelty in foundational physics.

The very place of thermodynamics in the panoply of fundamental physical theories seems very peculiar. Most of foundational physics consists either of basic kinematic and dynamic theories, such as classical, relativistic, or quantum dynamics with their associated underlying spacetimes, or of the laws governing the constitution of the basic substances of nature, such as the force laws governing the interactions of material particles or the basic field laws, such as those of electrodynamics. But thermodynamics does not fit into this scheme in any clear way. Yet its domain is universal. No matter what the constitution of a system in question—molecules, atoms, or electromagnetic radiation fields, say—and no matter what the appropriate dynamics that governs the system—classical, relativistic, or quantum mechanical—the basic concepts of temperature and entropy are applicable, and the basic notions of a time-asymmetrical tendency toward a spontaneously achievable equilibrium state apply.

2. FROM KINETIC THEORY TO STATISTICAL MECHANICS

By the nineteenth century it was widely accepted that heat was some kind of internal energy of a system. But in what form did this energy appear? Early suggestions by John Herepath and John Waterston that for gases the system could be described as a multitude of independently moving molecules interacting only by collisions were ignored. But this proposal gained traction when given new life by August Krönig. Clausius further advanced the claim by showing how collisions could explain slow diffusion, introducing the important notion of mean free path.

It was at the hands of Ludwig Boltzmann and Maxwell that the theory achieved its developed form. Maxwell proposed a plausible velocity distribution function for the molecules in the equilibrium condition. Both Maxwell and Boltzmann developed equations describing how a dilute gas would approach equilibrium from a nonequilibrium condition. This kinetic equation was subject to severe criticism, however.

A theorem of Boltzmann seemed to show that a system not in equilibrium would necessarily approach equilibrium in the future time direction and then stay in the equilibrium condition. But the underlying dynamics was time-reversible invariant, leading to the objection from reversibility, and the system was subject to a recurrence theorem of Henri Poincaré leading to the objection from recurrence. It was in the light of these objections that both Maxwell and Boltzmann introduced probabilistic notions into the theory. Maxwell, invoking the possibility of a "demon" who could subvert the second law by controlling the behavior of individual molecules, argued that the second law of thermodynamics was merely "probabilistic" in nature. But even the invocation of probabilities failed to answer the objections to Boltzmann's result, since one could still have time-symmetrical probabilities. Boltzmann's final answer invoked cosmology, the first version of an anthropic principle, and the claim that what was meant by the future direction of time was the direction in which, locally, entropy was showing a probabilistic increase.

Maxwell and Boltzmann developed a method for dealing with equilibrium that became the core of statistical mechanics. Fix the constraints on an equilibrium system, such as volume of container and total energy of the system. Look at the ensemble of all possible microstates of the molecules allowed by these constraints. Apply a probability distribution over this collection of possible microstates. Compute average values of quantities definable from the microstates of the system (phase averages). Identify these mean values with the observed macrofeatures of the equilibrium state of the system. This procedure was formalized by J. Willard Gibbs (microcanonical ensemble) and generalized by him for systems not thermally isolated but in contact with a constant temperature heat reservoir (canonical ensemble) and for a system allowing chemical interactions (grand canonical ensemble).

3. EQUILIBRIUM THEORY

A full understanding of equilibrium would require placing it in the general nonequilibrium theory and accounting for its status as the "end state" of spontaneous evolution. A great deal of effort, however, has gone into a program that treats equilibrium from another perspective. The standard procedure for calculating equilibrium values is to take them as mean values of phase functions calculated using the standard probability measure over the possible microstates of the system.

Much effort has gone into justifying the use of average values by arguing that for systems with vast numbers of degrees of freedom (the many molecules in a volume of gas) and for the kind of phase functions used in calculating mean values to identify with macroscopic features, the mean values so calculated will agree with the overwhelmingly most probable values of the phase functions using the same probability measure. Here deft applications of the laws of large numbers from probability theory, along with the use of fundamental characteristics of the molecular interactions, support the claim that means and most probable values may be taken to coincide.

But the deeper question is: Why adopt the standard probability distribution over the microstates to characterize equilibrium? One part of an answer goes like this: Assume that the equation governing nonequilibrium is correct. Then we can explore how any given probability distribution over the microstates will evolve in time by following the microevolution of each particular system in the collection over which the probability is defined. Use Boltzmann's theorem, the famous H-theorem, to show that the standard probability distribution is one that is stationary in time. Then argue that since equilibrium is, by definition, an unchanging macroscopic state, this shows that the standard probability distribution is the one to use if one is going to calculate macro values with it.

But this leaves open the following question: Is the standard probability distribution the only probability distribution that is time invariant? Trying to show that it is so is the subject of ergodic theory. Boltzmann was able to show the unique stationarity of the standard distribution by invoking a posit that a system started in any microstate would eventually move through all possible microstates. But that posit is provably false.

John Von Neumann and George Birkhoff were able to show that if a system met a certain dynamic condition (metric indecomposability), then a weaker version of so-called ergodicity could be demonstrated. That idealized systems such as hard spheres in a box interacting only by elastic collisions met the von Neumann–Birkhoff condition was proved much later by Yakov Sinai. Crudely, what the ergodic results show is that if a system has no hidden global constants of motion (and Sinai showed the hard sphere system did not), then the only probability distribution that assigned zero probability to the sets of conditions assigned zero probability by the standard distribution and that was invariant in time was the standard distribution.

Even within its own framework, ergodic theory as a rationale for equilibrium theory leaves open many mathematical, physical, and conceptual issues. From a philosophical perspective it is interesting for two main reasons. First, it offers a rationale for a probability distribution that is neither empirical nor founded solely on some a prioristic principle of uniformity or indifference. Second, the kind of explanation it offers is "transcendental" in nature. It does not explain why there are equilibrium states. It assumes that there are such time-invariant states and then, given a raft of assumptions about how macrostates are related to probabilities over microstates, it fixes the appropriateness of the probability from the assumed time invariance of the macrostate.

Such transcendental explanations in physics appear elsewhere. For example, one can derive features of the crystallization transformation in materials (or the transformation to a ferromagnetic state) not by giving a full nonequilibrium, dynamic explanation of the transition but by *assuming* that such a transition, one that takes a system from merely local order to infinitely extended global order, exists and then inferring the structure of the transition from that assumption.

4. NONEQUILIBRIUM THEORY

One major advance in the theory of nonequilibrium was the understanding of how Boltzmann was apparently able to derive a time-asymmetrical kinetic equation from time-symmetrical underlying dynamics. The solution is found in a posit Boltzmann made about collisions of the molecules. By assuming such collisions were random prior to interaction but not after interaction, the hypothesis about collision numbers, Boltzmann put time asymmetry into his derivation.

Further understanding came through an interpretation of the meaning of the Boltzmann equation due to Paul Ehrenfest and Tatiana Ehrenfest. Take a collection of systems prepared in a nonequilibrium condition. Watch them evolve over time. Understand the approach to equilibrium as being the fact that at future time instants the overwhelming majority of the systems would be found in states asymptotically closer to the equilibrium condition. The solution curve of the Boltzmann equation was then taken to describe this "concentration curve" of the evolving ensemble. Such an interpretation was consistent with the Poincaré recurrence result.

It was also consistent with a time-symmetrical understanding of the theory due to Boltzmann in his last publications. Here one viewed an isolated system as almost always at or near equilibrium, with excursions to nonequilibrium conditions that were rarer as the deviation from equilibrium was larger. Boltzmann reconciled this new time-symmetrical interpretation with our experienced time-asymmetrical thermodynamic world by cosmological arguments we will review shortly.

This interpretation avoids the paradoxes that were used to criticize the original understanding of the kinetic equation. But is there any reason to believe that realistic systems will show the kind of behavior this Boltzmann-Ehrenfest interpretation demands of a collection of systems? Here generalizations of ergodic theory have been employed to justify this picture of ensemble behavior.

Assume an idealized system, such as hard spheres in a box. Try to show that an ensemble corresponding to nonequilibrium will show a kind of approach to equilibrium. Here one uses a conception due to Gibbs. Gibbs described the mixing of insoluble black ink into water.

After a while we have an apparently uniformly gray mixture. But close inspection shows us that any point in the fluid is still black or transparent. The insoluble ink *mixes* into the water but does not dissolve in it. Similarly, what one tries to show is that the nonequilibrium ensemble will approach an equilibrium ensemble in a "coarse-grained" sense, since one can prove from the dynamics that it can never approach the equilibrium ensemble in a fine-grained sense analogous to the ink dissolving in the water.

A series of important mathematical results back up these claims. By using the randomizing aspect of collisions, one can show that ensemble evolution is "mixing" or even more strongly that ensemble evolutions are "K systems" or "Bernoulli systems." A significant aim of these mathematical programs is to eliminate a deep problem with Boltzmann's derivation of his kinetic equation. Boltzmann assumed that one could apply a "rerandomization" postulate for the conditions of the molecules at every moment of time. This is implicit in his hypothesis with regard to collision numbers. But the legitimacy of such a posit, and even its consistency with the dynamic laws, remains problematic. The mixing type results eschew any such rerandomization posits. One must be very careful, however, in using these mathematical results to claim a full understanding of the approach to equilibrium of actual physical systems. First, there is the highly idealized nature of the systems treated in the theory. More importantly, the mathematical results are often couched in such terms as what happens "in the limit as time goes to infinity." Results that would get us an understanding of the finite time results for real systems are much harder to obtain.

Finally, there is the fact that all of these results are symmetrical in time, since they are derived from the constitution of the system and the time-symmetrical underlying dynamics. The old questions about the origin of time asymmetry remain. As Nikolai Krylov pointed out, one could only obtain the desired time asymmetrical and finite time picture by positing appropriate initial ensembles with appropriate uniform probability distributions over initial conditions within them and refusing to allow the legitimacy of such posits as characterizing the final states of systems.

It is important to note that the approach to understanding nonequilibrium evolution through the ideas of mixing and a coarse-grained approach of a nonequilibrium ensemble to an equilibrium ensemble is not the only proposal on the table. Oscar Lanford and others have explored a rigorous derivation of the Boltzmann equation in which, in a special limit, one can prove that for a nonequilibrium ensemble and a reasonable probability distribution over it, the systems in the ensemble will, at least for a short time, move closer to equilibrium with probability one. This is quite a different probabilistic justification of the equation than those that understand the solution of the Boltzmann equation as merely tracking the concentration curve of the ensemble evolution. Some problems with this alternative approach are that it seems only applicable to dilute gases and the proved approach to equilibrium is only for miniscule times. As usual the issue of the origin of time asymmetry of thermodynamics remains unsolved by this approach in itself, since, once again, the formal results are all obtained from the underlying time-symmetrical dynamics.

The kinds of probabilistic explanations one finds in nonequilibrium theory are more in accordance with familiar philosophical models of statistical explanation than are the "transcendental" explanations of equilibrium theory. The core problems of nonequilibrium theory are the crucial issues centering around the legitimacy of the probabilistic posits needed over initial ensembles to obtain the realistic, time-asymmetrical behavior familiar to us from thermodynamics.

5. COSMOLOGY AND QUANTUM MECHANICS

Boltzmann was the first to introduce cosmology into the foundations of statistical mechanics. In his final version of statistical mechanics, systems were viewed time symmetrically as being almost always near equilibrium with occasional excursions to nonequilibrium states. But our world is one far from equilibrium and is time asymmetrical in its thermal behavior.

Starting from a suggestion by his assistant, Scheutz, Boltzmann offered the following story: The cosmos is, as a whole, near equilibrium. Finite regions of it can fluctuate into a nonequilibrium condition for finite periods of time. We are in such a region. Indeed we could only exist as sentient beings in a nonequilibrium fragment of the universe, since only there could energy flows exist to sustain complex, functioning organisms. In any nonequilibrium region, it is overwhelmingly likely that the region will be further from equilibrium in one time direction and closer in the other time direction rather than being further from equilibrium or closer to equilibrium in both time directions. But why do we find ourselves in a region that is getting closer to equilibrium in the *future* time direction? Because the very notion of what counts as future and what counts as past is grounded in entropic features of the world, just as what counts as up and what counts as down are founded on the local direction of the gravitational force.

With the development of modern observational cosmology, Boltzmann's picture fell into abeyance. Instead, the most common modern proposal is to found the entropic asymmetry of the world on some profound asymmetry in time of our universe on the cosmic scale. One proposal, not generally accepted, is to argue that it is the expansion of our universe from an initial Big Bang singularity that is responsible for entropic asymmetry. But if our cosmos were contracting, would entropic changes in local systems go the opposite way in time from the way they go now? Not according to prevalent opinion.

A more popular proposal is to argue that our universe's nonequilibrium and its time-asymmetrical approach to equilibrium are founded on the fact that the initial state of the universe is one of profoundly low entropy. On many of these models, whereas matter "just after" the Big Bang was at equilibrium, the uniformity of space provided a gigantic source of low entropy that could be called on, driving the matter to a new low-entropy condition. Crudely, matter originally at a uniform temperature and smoothly distributed in space changes to matter condensed into hot suns radiating out into cold space.

The question immediately arises: If high entropy is overwhelmingly probable, why is the Big Bang state one of such low entropy? Here answers range from adding new law-like constraints on singular conditions like the Big Bang to positing many universes of which ours is a rare exception (reanimating a new version of Boltzmann's argument) to denying that it makes sense to talk of the probability of some cosmic condition, such as the entropy of the Big Bang.

A profound issue in the use of this cosmology to resolve the time asymmetry problem of statistical mechanics is this: Perhaps positing initial low entropy, when combined with the usual time-symmetrical probabilistic posits, might give us an explanation of why the universe is not in equilibrium and why the cosmos as a whole shows time-asymmetrical entropic increase. But the real puzzle of statistical mechanics is not with the entropic change of the universe as a whole. The second law deals with the panoply of small systems in the universe that are in temporary thermal isolation from their environments—what Hans Reichenbach called "branch systems." (Reichenbach, 1956) It is the fact that these systems show a parallelism in time of their entropic change, with higher entropy being in and perhaps defining the future time direction, that we want to understand. Our theory without cosmology obtains this result by imposing time-asymmetrical boundary conditions on these systems. Can we do without these if we accept the time asymmetry of the cosmos?

Reichenbach was well aware of the difficulties here. Simply positing that branch systems will show an entropic increase that parallels that of the cosmos seems ad hoc. It is also hard to think of time-symmetrical probabilistic postulates that might be added to the cosmic hypothesis to get the results we want for branch systems. If we were talking about a partitioning of the cosmos into eternally isolated subcomponents, then some such time-symmetrical addition might do the trick, but we are talking about short-lived temporarily isolated systems with wildly varying past histories of causal interaction with their environments.

From the early days of statistical mechanics one proposal has always been to seek the time asymmetry of thermodynamics in some time asymmetry of the underlying dynamic laws. Early versions of this posited some kind of universal "friction," but this seemed dubious in the light of the demonstrable time asymmetry of classical mechanics and the usual time-symmetrical fundamental forces. There are other notable time asymmetries in physics, such as the fact that accelerated charged particles emit coherent outbound electromagnetic radiation, but we do not see such coherent radiation, "coming in from infinity," converging on charged particles.

The usual but not universal consensus has been that switching to quantum mechanics as the fundamental dynamic theory does not change the picture. The basic dynamic laws of quantum mechanics are still time symmetrical in their nature. "Measurements" in quantum mechanics are usually construed as time asymmetrical. Systems have their wave functions "collapse" into eigenstates of measured quantities after, and not before, measurements take place. The measurement problem in quantum mechanics remains of course highly controversial. Some accounts of measurement's apparent time irreversibility rely on explanations that presuppose thermal irreversibility. A similar view takes the apparent time asymmetry of radiation processes, with outbound coherent radiation from accelerated charges but no inbound coherent radiation that accelerated charges, as, again, thermodynamic and statistical mechanical in origin.

A recent interesting proposal suggests, once again, supplementing the cosmological posit of initial low entropy with an underlying asymmetrical fundamental dynamics to generate the time-asymmetrical parallelism of branch systems. One proposal to solve the measurement problem in quantum mechanics is to posit a stochastic interference in the evolution of systems from a time-asymmetrical probabilistic cause at a deeper level than that described by quantum mechanics. This proposal is often called the GRW approach after three of its initiators, Giancarlo Ghirardi, Alberto Rimini, and Tulio Weber. Recently, David Z. Albert and Barry Loewer have suggested that the time-asymmetrical GRW stochastic impulses on systems can also be invoked to account for the time asymmetry of branch systems. The details of this proposal remain a work in progress. And of course the belief in a GRW-like process as accounting for quantum measurement remains a minority view.

6. THE REDUCTION OF THERMODYNAMICS TO STATISTICAL MECHANICS

Statistical mechanics is supposed to give an explanatory account of the success of the older thermodynamic theory. To what extent can the relationship between the two disciplines be fit into a neat model of the older theory "reducing" to the newer? Such traditional reductions, with allowances for numerous subtleties, have the entities and features posited by the older theory identified with those posited by the newer and the law-like regularities of the older theory derivable from those of the newer given the appropriate identification of entities and properties.

One feature that stands out as blocking any simpleminded reductive account between the two theories is the fact that traditional thermodynamics states its laws in nonprobabilisitic terms, whereas the results of statistical mechanics are always framed in terms of probabilities. One ingenious resolution to this is the development of a statistical thermodynamics that, resting on a few simple principles, imports into thermodynamics itself probabilistic notions.

The connections between statistical mechanics and thermodynamics are also made subtle by the need to invoke various limits (of numbers of degrees of freedom in the "thermodynamic limit," for example) to derive the needed quasi-thermodynamic results from statistical mechanics. This opens up an area for exploring the many important and subtle ways such limiting processes can connect concepts and laws of theories at two different levels.

Another complication in the relation between the two theories is the way a single concept in thermodynamics splits into a whole family of concepts in statistical mechanics. The single concept of "entropy" in thermodynamics, for example, is associated in statistical mechanics with Boltzmann's entropy, Gibbs's fine-grained entropy, and Gibbs's coarse-grained entropy for different descriptive and explanatory purposes in statistical mechanics. Gibbs himself was careful to speak of "thermodynamic analogies" in statistical mechanics rather than of identifications of thermodynamic and statistical mechanical features.

There is an interesting sense in which thermodynamics characterizes the world with great independence from any specific views about the substantial nature of the systems being so characterized. Temperature and entropy can, for example, be instanced in systems as diverse as gases consisting of molecules in a box and trapped electromagnetic radiation. Here a useful analogy can be made to the claim of functionalism in philosophy of mind that tries to argue that mental states are functionally defined and could be instanced in a wide variety of underlying physical systems.

7. THE DIRECTION OF TIME

We noted earlier that an important part of Boltzmann's final cosmological version of his theory was the claim that in a region of the universe that had lower entropy in one time direction and higher entropy in the other time direction, the inhabitants of that region would count as the future time direction the time direction in which entropy was increasing. What would be needed to make such a claim plausible? It could not be that we normally inferred the future direction of time by making observations on the entropic changes of systems. After all, even in the absence of any external information, we know of two mental events, say, which is earlier and which later.

Insight into these issues can be obtained by looking at the analogy that Boltzmann draws between the past-future case and the case of up and down. We distinguish between the upward and downward spatial directions in our vicinity. We can do this using objects (which way do rocks fall) and even without external observations. Our current theory tells us, however, that what we meant by "down" all along was just the local direction of the gravitational force. We are perfectly happy to take it that "down" directions are antiparallel for antipodean dwellers on earth, and we are happy to concede that in regions of the universe where there is no gravitational force, no direction is down. All of this is true despite the fact that someone can tell which direction is down without having the least comprehension of the notion of gravitational force.

Similarly, argues Boltzmann, in regions of the cosmos in which entropic change is counterdirected in time, past and future will be in opposite time directions. And in an equilibrium region of the universe there will be no distinction between past and future, although there will still be two opposite directions in time.

On the other hand, consider our distinction between left- and right-handed objects. There appears to be a deep, law-like physical asymmetry between left and right orientations in the weak interactions of elementary particles (parity nonconservation results). But it seems implausible to us to say that the left-right distinction for us is somehow grounded on these physical asymmetries. What is the difference in the cases of up and down and left and right?

It is that in the up-down case all of the usual means by which we ordinarily distinguish up from down (the direction in which rocks fall as down, for example) are accounted for by reference to the gravitational force. That force also explains the behavior of the fluids in our inner ears that give us our immediate noninferential grasp of which direction is down. Such an explanatory account of our left-right distinction in terms of the physical asymmetries of weak interactions is not forthcoming.

Following this line of reasoning, what would it take to make the Boltzmann claim about the past-future distinction plausible? The Boltzmann thesis could only be plausible if we could offer explanations, framed in terms of the asymmetry in time of entropic change, of all the phenomena by which we ordinarily distinguish past and future. The explanatory account should also explain how it is that we can have a noninferential, direct awareness of the time order of events in our immediate apprehension.

How do we distinguish past from future? Well, we have records of the past but not of the future and memories of the past but not of the future. Somehow the entropic account should explain this differential epistemic access to what has happened as opposed to what will happen. We take it that causation goes from past to future. How does the entropic asymmetry account for that intuitive distinction between past and future? We think of the past as "fixed" with a determinate structure and as unchangeable. What does entropy have to do with such an intuition? Why is it that we view the future with anxiety but the past with regret? Again, this seems remote from anything having to do with the fact that we can stir cream into coffee but cannot stir it out.

There have been a number of attempts at filling out the Boltzmann claim to make it more plausible. Such programs usually work by taking one of the intuitive distinctions between past and future as primitive and trying to derive the other distinctions from that basic one. Then one will try to show how the entropic asymmetry of the world gives a full explanatory account of the primitive intuitive distinction.

For example, one might take as basic the fact that we have records and memories of the past but not of the future. Then one could argue that our idea that the past is determinate and the future a realm of possibilities comes from this asymmetrical cognitive access to past and future. One might also argue that our idea of causation as going from past to future hinges on this asymmetry of knowledge. Or one might argue that our notion of causation is grounded on that of counterfactuals (if the cause had not occurred the effect would not have occurred). One could then seek principles of evaluation of counterfactuals that rely on the entropic asymmetry to explain how we evaluate these in a manner that grounds time-asymmetrical causation.

All one can say at the present time is that the search for such explanatory accounts of our intuitive past-future distinction resting on considerations of the entropic asymmetry in time of branch systems remains open. Behind all the ongoing programs of trying to fill out the Boltzmann claim to make it fully plausible is one fundamental consideration, however. Putting subtle asymmetries in time of elementary particle interactions to the side, the physical time asymmetry of systems, the fundamental time asymmetry of the world, does seem to be captured in the entropic time asymmetry. Broad physicalist considerations support the claim that an entropic account must, then, lie at the heart of any explanation of our notion of the direction of time.

Further Readings

An overall survey of the issues discussed in this article is Sklar 1993. For the history of statistical mechanics, see Brush 1976 and 1983. A useful collection of original works in the field is Brush 1965. A seminal work in statistical mechanics is Gibbs 1960. Ehrenfest and Ehrenfest 1959 set the stage for all future foundational work in statistical mechanics. On explanations in other areas of physics that are like the "transcendental" explanations of equilibrium theory, see Batterman 2002 and Bruce and Wallace 1989. For the importance of initial conditions on branch systems, Krylov 1979 is fundamental. On cosmology and statistical mechanics, Davies 1974, chap. 4, is a fine introduction. On the notion of "branch systems," see Reichenbach 1956. For the issue of radiation asymmetry and entropic asymmetry, see Davies 1974, chap. 5, and Frisch 2005, pt. 2. On the possibility of invoking the GRW approach to quantum measurement to ground time asymmetry in statistical mechanics, see Albert 2000, chap. 7.

References

Albert, D. 2000. *Time and Chance*. Cambridge, MA: Harvard University Press.
Batterman, R. 2002. *The Devil in the Details*. Oxford: Oxford University Press.
Bruce, A., and D. Wallace. 1989. "Critical Point Phenomena." In *The New Physics*, ed. P. Davies, 236–267. Cambridge: Cambridge University Press.
Brush, S. 1976. *The Kind of Motion We Call Heat*. Amsterdam: North-Holland.
Brush, S. 1983. *Statistical Physics and the Atomic Theory of Matter*. Princeton, NJ: Princeton University Press.
Brush, S. ed. 1965. *Kinetic Theory*. Oxford: Pergamon.
Davies, P. 1974. *The Physics of Time Asymmetry*. Berkeley: University of California Press.
Ehrenfest, P., and T. Ehrenfest. 1959. *The Conceptual Foundations of the Statistical Approach in Mechanics*. Ithaca, NY: Cornell University Press.
Frisch, M. 2005. *Inconsistency, Asymmetry, and Non-Locality*. Oxford: Oxford University Press.
Gibbs, J. 1960. *Elementary Principles in Statistical Mechanics*. New York: Dover.
Krylov, N. 1979. *Works on the Foundations of Statistical Physics*. Princeton, NJ: Princeton University Press.
Reichenbach, H. 1956. *The Direction of Time*. Berkeley: University of California Press.
Sklar, L. 1993. *Physics and Chance*. Cambridge: Cambridge University Press.

INDEX

Figures and notes are indicated by f and n following the page number.

Absoluteness, 198–199
Abstract algebra, 240–241
Acceleration, 219
Accidental facts, 63, 69–70, 73–76
Achinstein, Peter, 52
Actualism, 263–264
Adaptation requirement in evolutionary models, 164
Albert, David Z., 279
Algebraic framework for QM∞, 229, 243–250
Algebraic imperialism, 254–255
The Analysis of Matter (Russell), 129
Antecedent conditions, 16
Antiessentialism, 110
Antirealism, 108. *See also* Instrumentalism
Apologetic imperialism, 255–258
Arageorgis, Aristidis, 254
Arbitrariness of coordinate systems, 219–220
Aristotelian tradition, 178
Armstrong, D. M., 76, 80n
Asymmetry: of causation, 19, 33–34; of knowledge, 283; of time, 278, 283
Authority requirement in revolutionary models, 171

Bayes, Thomas, 171
Bayesian models, 170–171
Bennett, Jonathan, 65
Bernoulli systems, 275
Beth, E. W., 150
Big Bang cosmology, 190, 212–213, 213f, 219, 277–278
Bildtheorie, 129
Biodiversity, 87
Biology, 10, 31, 75, 84
Blackburn, Simon, 103, 105–108, 105n, 115–116, 116n, 122
Black holes, 212
Bohr, Niels, 151
Bold imperialism, 258–260
Boltzmann, Ludwig, 271, 272, 274–277, 280–282
Boyd, Richard, 95
Boyle's law, 83, 84, 85, 86, 87
Branch systems, 278
Bridgman, Percy, 99

C*-algebras, 244–246
Canonical anticommutation relations (CARs), 229, 250–254
Canonical commutation relations (CCRs), 229, 250–254

INDEX

Canonical coordinates, 231
Cardinality constraint, 138
Carnap, Rudolph, 99–100, 127, 130–133, 143–144, 152, 173
Carnot, Nicolas Léonard Sadi, 269
Carroll, John W., 65
CARs (canonical anticommutation relations), 229, 250–254
Cartwright, Nancy, 21n, 81n, 84, 97
Categorical properties, 76
Category theory, 157n
Cauchy surfaces, 264
Causal abundance, 214–215, 214f, 216f
Causal interaction, 22
Causal isolation, 212–214, 213f
Causality: asymmetry of, 19, 33–34; DN/IS model, 14–15; and entanglement of spacetime, 209–212, 210–211f; and laws of nature, 15; probabilistic explanation, 48–50; role of, 32–34
Causal/mechanical (CM) model, 22–25
Causal theory of time, 216–217
CCR/CAR algebra, 229
CCRs (canonical commutation relations), 229, 250–254
Central limit theorem, 60
Ceteris paribus qualification, 83–87
Chakravartty, Anjan, 151
Change debate, 53–56
Chemistry, 86–87, 178
Chisholm, Roderick M., 65
Classical vs. relativistic spacetimes, 192–194, 193f, 222–225, 222–223f, 225f
Clausius, Rudolf, 270
Clifton, Robert, 260
CM (causal/mechanical) model, 22–25
Coffa, Alberto, 48
Cognitive evolution, 164–165
Commutant, 247
Competing interests requirement in revolutionary models, 172
Competition requirement in evolutionary models, 164, 166–167
Concentration curve, 274
"A Confutation of Convergent Realism" (Laudan), 147
Conservation principle, 202–203

Constructive empiricism, 104, 150
Constructive realism, 150
Contingent necessities, 64, 72
Contingent truths, 64
Conventionalism, 128, 197
Conventionality of simultaneity, 196–198
Coordinate systems, 219–220
Copernicus, 179
Correspondence theory, 95, 131
Cosmology, 212, 276–279
Coulomb's law, 78, 79
Counterfactual conditionals, 66
Counterfactuals, 13, 64–69
Course-grained entropy, 280
Craig, William, 100
Craigification of theoretical content, 241–242
Cumulativism, 174, 178, 179

Darwin, Charles, 164
Darwinian evolutionary models, 164–167
Davidsonian-Tarskian referential semantics, 111
Deductive-nomological model. *See* DN model
Deductive-nomological-probabilistic account, 52
Deductive systematizations, 36
Deflationary arguments, 43, 50, 96n
Densities, 42, 203–204
Dependence, 19, 21, 28
Dephlogisticated air, 118–119
Determinateness of future, 195–196, 196f
Determinism, 56–61, 243
Deviant models, 139–142, 142–143, 145
Dewey, John, 99
DHR (Doplicher, Haag, Roberts) selection theory, 264–265
Disjointness, 249–250
Disjunctive property, 76
Dispositionalism, 50, 56
Diversity requirement in evolutionary models, 164, 169
DN (deductive-nomological) model: context for, 14–15; counterexamples, 16–19, 32, 48; defined, 10–11; fate

INDEX

of, 30–32; motivations for, 15–16; necessity, 16–19; sufficiency, 16–19
Double commutant, 247
Dowe, Philip, 23
Dretske, Fred I., 76
DTA account, 76, 80
Dynamical automorphism, 263
Dynamicism, 264

Earman, John, 202
Economics, 75
Egalitarianism, 53–56
Ehrenfest, Paul, 274
Ehrenfest, Tatiana, 274
Einstein, Albert, 148, 185, 188, 196–198, 208–209, 209n, 220. *See also* Relativity
Electrical theory, 178
Elite property instantiations, 76
Elitism, 53–56
Ellis, Brian, 99
Energy: CM (causal/mechanical) model, 22, 23; entanglement of spacetime and matter, 206–207
Entanglement of space and time, 188–198; classical vs. relativistic spacetimes, 192–194, 193f, 222–225, 222–223f, 225f; conventionality of simultaneity, 196–198; determinateness of future, 195–196, 196f; infinitesimal neighborhoods of events, 190–192, 191–192f; relativity of simultaneity, 188–189, 189f, 222–225, 222–223f; robustness, 190; time as fourth dimension, 194–195
Entanglement of spacetime and causation, 208–217; causal abundance, 214–215, 214f, 216f; causal isolation, 212–214, 213f; causal structure, 209–211, 210f; causal theory of time, 216–217; relation to causality, 211–212, 211f; speed of light, 208–209
Entanglement of spacetime and matter, 198–208; absoluteness, 198–199; densities, 203–204; energy, 206–207; force, 206–207; geometry, 208; gravitational energy and momentum of extended systems, 205–206; gravitational field energy momentum, 202–203; hole argument, 200–201, 200f; manifold of events and metrical field, 201–202; momentum, 206–207; novelty, 202; substantivalism, 199–200, 199f
Entity realism, 97
Entropy, 270, 280
Epistemic instrumentalism, 98, 103
Epistemic relativity, 47–48
Epistemological naturalism, 141
Equilibrium theory, 253, 272–273
Ergodic theory, 273
Essentialism, 81–82, 82n
Euclidean geometry, 185, 219, 253
Everything Must Go (Ladyman & Ross), 152
Evolutionary models, 164–171
Expectability, 51
Explanation (scientific), 9–39; causation's role in, 32–34; CM (causal/mechanical) model, 22–25; DN (deductive-nomological) model, 10–11, 14–19, 30–32; functional explanation, 14–15; IS (inductive-statistical) model, 11–12, 14–19; role of laws, 12–13, 34–37; SR (statistical relevance) model, 19–21; unificationist models, 26–30. *See also* Probabilistic explanation
Extreme egalitarianism, 55
Extreme elitism, 53

Fell's theorem, 256
Fetzer, James, 48
Fine, Arthur, 108–111, 112n, 113n, 114–115, 114n, 116n
Fine-grained entropy, 280
First-order domains, 134, 135
First-order logic, 134
First-order semantics, 136
Fit of laws, 77
Folia, 249–250
Force, 172, 206–207
Fragility, 49
French, Steven, 150
Frequency theory, 57
Friedman, Michael, 26, 28
Full structures, 134
Functionalism, 14–15, 280

INDEX

Galilean strategy, 96
General covariance principle, 200, 219
Generalizations, 34–35, 36, 170
Genetic mutation, 165
Geometry: entanglement of spacetime and matter, 208; relativity of, 220; of spacetime, 198, 206–207, 214
Ghirardi, Giancarlo, 279
Gibbs, J. Willard, 271, 274, 280
Gibbs state, 253
Giere, Ronald N., 84, 150
Gimbel, Steven, 53
Gleason's theorem, 234, 248
Gluck, Stuart, 53
GNS representations, 247–249, 251, $251n$, 261–262
Goodman, Nelson, 65
Gravitational energy and momentum of extended systems, 205–206
Gravitational field energy momentum, 202–203
Gresham's law, 83, 85, 86, 87

Haag, Rudolf, 241
Hacking, Ian, 97
Hadamardism, 264
Halvorson, Hans, 231, 260
Hamilton, William Rowan, 231
Hamilton's equation, 27, 217, 231
Hardy-Weinberg equation, 84
Heat energy, 269–270
Heliocentric model, 179
Helmholtz, Hermann von, 128, 129
Hempel, Carl, 10–16, 44–48, 51, 53, 57, 75, 100, 133, 241–242
Henkin structures, 134
Herepath, John, 270
Hertz, Heinrich, 128, 129
Hilbert space conservatism, 250–254, 260, 263
Hilbert space representation, 229, 230
Hilbert space vector norm, 245
Hitchcock, C., 24
Hole argument, 200–201, 200f
Holism, $77n$
Hooke's law, 75
Horwich, Paul, 65, 103

H-theorem, 272
Humean accounts, $78n$, 79
Humphreys, Paul, 50, 51

Incommensurability, 173
Independence, 19, 21
Indeterminancy of translation, 140
Indexicality, 153, 156
Inductive confirmation, 73–76
Inductive-statistical model. *See* IS model
Inexact sciences, 64, 83–89
Inference tickets, 100, 101
Infinitesimal neighborhoods of events, 190–192, 191–192f, $192n$
Informativeness of laws, 77
Inheritance requirement in evolutionary models, 164
Instrumentalism, 94–126; and quietism, 108–116; realism vs., 94–108; reclaiming, 116–125; and structuralism, 133–136
Intentionality, 153
Internal realism, $96n$, 99, $102n$
Intertheory relations, 157
Irreducible propensity, 57
IS (inductive-statistical) model: context for, 14–15; counterexamples, 16–19; defined, 11–12; motivations for, 15–16; necessity, 16–19; and nomic expectability, 44–48; objections, 47–48; and probabilistic explanation, 57; radioactive decay, 49; sufficiency, 16–19

Jackson, Frank, 59, 65
James, William, 99
Jeffrey, Richard, 48
Johnson, Samuel, $113n$

Kadison, Richard V., $262n$
Kant, Immanuel, 129, 185. *See also* Neo-Kantianism
Kastler, Daniel, 241
Kay, Bernard S., 252
Kepler's first law of planetary motion, $64n$
Kerr solution, 31
Ketland, Jeffrey, 146
Killing field, $205n$

INDEX

Kinematics, 229
Kinetic theory, 270–272
Kitcher, Philip, 26–27, 28, 36n, 95
Klein-Gordon equation, 35, 239, 240, 252, 257
KMS (Kubo, Martin, & Schwinger) equilibrium, 253
Know-nothingism, 113, 114, 114n
Kretschmann, Erich, 220
Krönig, August, 271
Krylov, Nikolai, 275
K systems, 275
Kuhn, Thomas, 147, 173–176, 177–178

Ladyman, James, 150, 152
Lagrange's equation, 27, 217
Lanford, Oscar, 276
Lange, Marc, 2, 63
Laudan, Larry, 106, 147, 157
Lavoisier, Antoine, 178
Laws of nature, 63–93; as collective, 72; and counterfactuals, 64–69; defining, 76–83; in DN model, 11; and evolutionary change, 168, 169; explanation role of, 12–13, 34–37; inductive confirmation, 73–76; and inexact sciences, 83–89; nomic necessity, 69–73; stability, 69–73; statistical, 11–12, 44–48
Leplin, Jarrett, 3, 163
Lewis, David, 68, 76, 76n, 77n, 78, 78n, 80–81, 142–147
Light cone structure, 210, 210f, 211
Local spacetime theories, 202
Loewer, Barry, 279
Logic, 127, 134
Logical empiricism, 100, 127, 173
Logical positivism, 99, 100
Long-term potentiation (LTP), 29
Lorentz normal coordinate system, 192n

Mach, Ernst, 129, 185, 185n, 221
Mackie, J. L., 65
Malament-Hogarth spacetime, 215, 216f
Manifold of events, 195, 199, 201–202
Manifold substantivalism, 200, 202
Mathematics, philosophy of, 138

Maximal specificity requirement, 47
Maxwell, James Clark, 26, 148, 188, 271
McKenzie, Kerry, 3, 127
Mechanical systems, 23, 25
Melia, Joseph, 145–146, 147, 154
Mendelian genetics, 35, 84
Metalaws, 66
"The Methodological Character of Theoretical Concepts" (Carnap), 130–133
Metrical field, 199, 201–202
Metric indecomposability, 273
Microevents, 86
Mill, John Stuart, 76
Mill-Ramsey-Lewis (MRL) theory, 13, 36, 36n, 76
Mini-spacetimes, 190–192, 192n, 222–225
Minkowski, Hermann, 188, 192n, 195, 199, 206, 207, 213n, 218, 219, 254
"Models and Reality" (Putnam), 137, 138
Moderate egalitarianism, 55
Moderate elitism, 54, 55–56
Modesty requirement, 186–187
Momentum: CM (causal/mechanical) model, 22, 23; entanglement of spacetime and matter, 206–207; gravitational energy and momentum of extended systems, 205–206; gravitational field energy momentum, 202–203
Monod, Jacques, 85n
Moore, G. E., 113n
MRL. See Mill-Ramsey-Lewis theory
Müger, Michael, 231
Mumford, Stephen, 81n
Musgrave, Alan, 111, 112, 113, 113n, 114, 114n

Nagel, Ernst, 100–101, 102
Naive regularity accounts, 78n
Naturalism, 80n, 141, 154–159
Natural kinds, 74–76
Natural laws. See Laws of nature
Natural Ontological Attitude (NOA), 108–116
Natural selection principle, 84, 84n

289

INDEX

Necessity requirement, 63
Neo-Kantianism, 128–130, 154, 155
Neutral nodes, 159
Newman, Maxwell, 129, 133, 154
Newtonian gravitational theory, 25, 204n
Newtonian mechanics, 117, 120, 121, 174, 178, 185, 217, 231–232
NOA (Natural Ontological Attitude), 108–116
Nomic expectability, 15, 44–48
Nomic necessity, 63, 69–73, 76, 81, 82
Nomic preservation, 67–69, 70, 79
Nonequilibrium theory, 271, 274–276
Nonuniqueness in QM∞, 236–241
"No-overlap" principle, 146
Nordström theory of gravitation, 209n
Norm convergence, 246
Norm limits, 259
Norm topology, 240, 245
Norton, John D., 4, 185, 202
"No-theory" conception of truth, 110
Novelty requirement, 186, 189, 191, 202

Observation language, 145
Observation predicates, 138
"On the Electrodynamics of Moving Bodies" (Einstein), 188
On the Origin of Species (Darwin), 164
Ontological "bracketing," 107
"Ontological Relativity" (Quine), 141
Ontology of space and time, 185–228; arbitrariness of coordinate systems, 219–220; entanglement of space and time, 188–198; entanglement of spacetime and causation, 208–217; entanglement of spacetime and matter, 198–208; modesty requirement, 186–187; novelty requirement, 186; realist requirement, 187; relational view of, 220–221; relativistic morals that founder, 217–221; relativity of geometry, 220; relativity of motion, 218–219; requirements, 186–188; robustness requirement, 187
Operationalism, 259
Ordinary quantum mechanics (QM), 229. *See also* Quantum mechanics
Orthogonality, 190n, 250

Paradoxes of perspective, 156, 158
Paramagnetism, 49
Partial interpretation, 136
Particle horizon, 213
Pauli spin observables, 233, 238
Peirce, C. S., 99
Pessimistic induction, 97, 102n, 155, 157
Pettit, Philip, 59
Phase structure, 253
Phlogistic chemistry, 178
Physical necessity, 63
Physics, 3–5, 185–284; DN model, 31; and instrumentalism, 120; ontology of space and time, 185–228; QM∞, 229–268; and scientific explanation, 10; spacetime in, 195–196; statistical mechanics, 269–284
Planck, Max, 128, 129
Poincaré, Henri, 128, 148, 220, 271, 274
Poisson bracket, 232
Polarization of system, 236, 238, 244
Pollock, John L., 65
Popper, Karl, 57
Positivism, 129, 173, 174
Pragmatism, 99
Predicates, 12–13, 138–139, 146
Priestly, Joseph, 118, 119
Primas, Hans, 263
Probabilistic explanation, 40–62; causal approaches, 48–50; determinism, 56–61; egalitarianism, 53–56; elitism, 53–56; IS (inductive-statistical) model, 44–48; and nomic expectability, 44–48; probabilistic relevance accounts, 50–53; SR (statistical relevance) model, 19–21; varieties of, 40–44; Probabilistic relevance accounts, 50–53
Projectable predicates, 12–13
Propensity theory, 57
Proxy functions, 141, 159
Pseudoprocesses, 22, 23–24
Pseudotensor, 204
Psillos, Stathis, 148
Psychology, 10
Putnam, Hilary, 95, 99, 137, 138, 138n, 140, 142
"Putnam's Paradox" (Lewis), 143

INDEX

QFTs. *See* Quantum field theories
QM∞, 4, 229–268; actualism, 263–264; and algebraic framework, 243–250; algebraic imperialism, 254–255; apologetic imperialism, 255–258; bold imperialism, 258–260; C*-algebras, 244–246; DHR selection theory, 264–265; disjointness, 249–250; dynamicism, 264; GNS representations and algebraic states, 247–249; Hadamardism, 264; Hilbert space conservatism, 250–254; interpreting, 241–243, 250–265; mixed strategies for interpreting, 260–261; nonuniqueness in, 236–241; quantizing, 231–233; quasi-equivalence, 249–250; representing CARs, 233; representing CCRs, 231–233; symmetry mongering, 262–263; tempered universalism, 261–265; uniqueness in ordinary QM, 233–235; universalism, 261–265; Von Neumann algebras, 246–247
QSM. *See* Quantum statistical mechanics
Qualitative predicates, 12–13, 77
Quantum field theories (QFTs), 230, 243, 252, 254
Quantum mechanics: and cosmology, 276–279; indeterministic aspect of, 40–41; measurement problem, 279; statistical mechanics, 276–279
Quantum statistical mechanics (QSM), 230, 253, 264
Quasi-equivalence, 249–250
Quietism, 108–116
Quine, W. V. O., 119–120, 140–142, 145, 156–158

Radical translation, 140, 141
Radioactive decay, 49
Railton, Peter, 52, 55, 58
Ramsey, Frank, 3, 76
"Ramseyan humility," 142–147
Ramsey sentence, 3, 131–136, 144, 146, 149, 154
Realism, 94–126; instrumentalism vs., 94–108; ontology of space and time, 187; and quietism, 108–116; and relativity, 187, 221; and spacetime substantivalism, 200; and structuralism, 133–136, 144–145
Reason, Truth, and History (Putnam), 138*n*
Regularity accounts, 78*n*
Reichenbach, Hans, 196–198, 216–217, 220, 278
Relativity: classical vs. relativistic spacetimes, 192–194, 193*f*, 222–225, 222–223*f*, 225*f*; general covariance of, 200, 219; of geometry, 220; of motion, 218–219; ontological requirements, 186–188; of simultaneity, 188–189, 189*f*, 222–225, 222–223*f*; Relativity postulate, 218; Relevance: explanatory, 60; probabilistic relevance accounts, 50–53; and scientific explanation, 17; and SR (statistical relevance) model, 19
Revolutionary models of theory change, 171–176, 179–180
Rimini, Alberto, 279
Ringrose, John R., 262*n*
Robertson-Walker spacetimes, 190, 213*n*, 217*n*, 225
Robinson, D. W., 263
Robustness requirement: entanglement of space and time, 190; ontology of space and time, 187; probabilistic explanation, 58, 60
Ross, Don, 152
Ruetsche, Laura, 4
Russell, Bertrand, 128, 129

Saatsi, Juha, 146, 147, 154
Salam, Abdus, 26
Salmon, Wesley, 19–21, 22, 24, 29, 51–52, 55, 57
Saunders, Simon, 3, 127
Schrödinger equation, 13, 25, 31, 35
Schwarzschild solution, 31, 190, 212
Scientific change. *See* Theory change
The Scientific Image (van Fraassen), 153
Scientific method, 1–3, 9–181; and instrumentalism, 94–126; laws of nature, 63–93; probabilistic explanation, 40–62; and quietism,

Scientific method (*Cont.*) 108–116; and realism, 94–108; scientific explanation, 9–39; structuralism, 127–162; theory change, 163–181
Scientific Representation (van Fraassen), 153
Scriven, Michael, 53–54
Second-order domains, 134, 135n, 143
Second-order interpretative structures, 134
Second-order quantification, 136, 144
Segal, Irving E., 254, 258–259, 261
Semantic approach, 99, 101, 150–153
Set theory, 127, 151, 156, 158
Sewell, Geoffrey L., 231
Simplicity of laws, 77
Simultaneity: ambiguity of, 197; conventionality of, 196–198; in mini-spacetimes, 222–225, 222–223f, 225f
Sinai, Yakov, 273
Single particle states, 240
Size debate, 53–56
Sklar, Lawrence, 5, 269
Sneed, Joseph, 150
Snell's law, 75
Social sciences, 31
Sophisticated regularity accounts, 78n
Spacetime. *See headings starting with* "Entanglement"
Speed of light: and simultaneity, 188; variability of, 208–209, 209n
Spin chain, 244–245
Spontaneity, 165
SR (statistical relevance) model, 19–21
Stability of laws of nature, 69–73
Stachel, John, 200
Stalnaker, Robert, 145
Stanford, P. Kyle, 2, 94, 120n
Statistical mechanics, 269–284; cosmology and quantum mechanics, 276–279; direction of time, 281–283; equilibrium theory, 272–273; kinetic theory, 270–272; nonequilibrium theory, 274–276; and probabilistic explanation, 41–42, 58; thermodynamics, 269–270, 279–280
Statistical relevance account, 51–52, 55

Stegmüller, Wolfgang, 150
Stein, Howard, 103
Stochastic interference, 279
Stone-von Neumann uniqueness theorem, 230, 234, 235, 236
Strawson, P. F., 65
Stress-energy tensor of matter, 203–204
Strevens, Michael, 2, 40, 54, 59–60
Structuralism, 127–162; Carnap's version, 130–133, 143–144; enforcement of, 137–139; and instrumentalism, 133–136; and Lewis's "Ramseyan humility," 142–147; and meaning, 140–142; naturalist approach, 154–159; neo-Kantianism, 128–130; and realism, 133–136; semantic approach, 150–153; and theory change, 147–149; Structural realism, 97, 128, 149, 152
"Structural Realism: The Best of Both Worlds?" (Worrall), 148
Substantivalism, 199–200, 199f
Summers, Stephen J., 257
Suppe, Frederick, 127, 128
Suppes, Patrick, 150
Swammerdam, Jan, 85n
Symmetry: and laws of nature, 66; and QM∞, 262–263; in spacetime coordinates, 195n; and structuralism, 158

Tachyons, 210
Tarski, Alfred, 158
Tempered universalism, 261–265
Theorem-style eliminibility, 100–101
"Theoretician's dilemma," 100, 133, 241
Theory change, 163–181; complexity of scientific change, 163; Darwinian evolutionary models, 164–167; evolutionary models, 164–171; model comparisons, 176–179; revolutionary models, 171–176, 179–180; and structuralism, 147–149
Theory choice, 167n
Thermodynamics: asymmetry of, 33; statistical mechanics, 269–270, 279–280; and time asymmetry, 278
Time: asymmetry of, 278, 283; direction of, 281–283; and quantum mechanics,

INDEX

194–195; and statistical mechanics, 281–283
Time travel, 214f, 215
Tooley, Michael, 76, 80n
Transcendentalism, 273, 276
Transitivity, 197
Trivialization, 135–136, 137
"Two Dogmas of Empiricism" (Quine), 140

Unconceived alternatives problem, 97–98
Underdetermination, 21n, 97
Unificationist models of explanation, 26–30, 36, 36n
Uninstantiated universals, 80n
Uniqueness in ordinary QM, 233–235
Unitary equivalence, 230, 235, 239, 255
Universalism, 261–265

Van Fraassen, Bas, 66n, 103–107, 105n, 122, 144, 150–153, 156, 158, 167n

Verificationism, 140
Von Neumann, John, 273
Von Neumann algebras, 246–247, 251, 260, 261

Wald, Robert M., 252, 264
Waterston, John, 270
Weak equivalence, 255, 256
Weber, Tulio, 279
Weinberg, Steven, 26
Weyl algebra, 240, 254, 257
Wigner, Eugene Paul, 66n
Wigner's theorem, 234, 236, 238
Wittgenstein, Ludwig, 128
Woodward, James, 1, 9, 24, 59
Word and Object (Quine), 140
Wormholes, 215
Worrall, John, 97, 128, 148, 149, 154

Zahar, Elie, 149